Oscar Harger

Report on the Marine Isopoda of New England and adjacent Waters

Oscar Harger

Report on the Marine Isopoda of New England and adjacent Waters

ISBN/EAN: 9783337141462

Printed in Europe, USA, Canada, Australia, Japan

Cover: Foto ©ninafisch / pixelio.de

More available books at **www.hansebooks.com**

REPORT ON THE MARINE ISOPODA OF NEW ENGLAND AND ADJACENT WATERS.

BY OSCAR HARGER.

[FROM THE REPORT OF THE UNITED STATES COMMISSIONER OF FISH AND FISHERIES, PART VI FOR 1878.]

XIV. REPORT ON THE MARINE ISOPODA OF NEW ENGLAND AND ADJACENT WATERS.

BY OSCAR HARGER.

The following paper includes the species of Isopoda at present known to inhabit the coast of New England and the adjacent regions, as far as Nova Scotia on the north and New Jersey on the south. These limits have been chosen from the fact that nearly all the marine collections of this order made by the Fish Commission have been from the New England coast, except those from the Nova Scotia coast in 1877, while the commission had its headquarters at Halifax. Previous to the work of the Fish Commission extensive collections had also been made, mostly by Professors A. E. Verrill and S. I. Smith, of Yale College, in the Bay of Fundy and at other places along the coast as far south as Great Egg Harbor, in the southern part of New Jersey. The collections thus obtained, and others in the museum of Yale College, have, through the kindness of Professor Verrill, been used in the preparation of this article. As there has not yet been sufficient opportunity for the study of the *Bopyridæ*, only a list of the known species of that family is included, and for this I am indebted to Professor S. I. Smith. The species of the remaining families are described at length, and nearly all figured in more or less detail in the plates accompanying the article. Throughout the article especial reference will be had to the Isopoda of our own coast, and many peculiarities of structure, not found in our genera, will be more or less completely disregarded. As the *Oniscidæ* are a terrestrial family, only a few species, found usually, or only, along the shore are here included.

ISOPODA.

This group is an order of Crustacea, so named from two Greek words, ἴσος, equal, and πούς, a foot, from the general similarity of the legs throughout, all being thoracic. The order belongs to the *Tetradecapoda*, "fourteen-footed," called also *Edriophthalma*, or "sessile-eyed" Crustacea. All of these terms, however, require modification when applied to the animals included in this order, since in the genus *Astacilla* the anterior pairs of legs are quite unlike the posterior, in *Gnathia* there are never more than twelve feet, or legs, in six pairs, and lastly in *Tanais* and its allies the eyes, when present, are not sessile, but articulated with the head, or stalked, as in the higher Crustacea. It may, however, be stated that the relations of the *Tanaidæ* with the rest of the order are remote, and it is perhaps doubtful whether they should be retained among the *Isopoda*, especially as this family differs from the rest of the order in its mode of respiration, as will be explained hereafter.

Although this order is not a large one its representatives are perhaps more widely distributed than in any other order of Crustacea. Every one is familiar with "sow-bugs" or "pill-bugs," which are found even in damp houses and in cellars, as well as under leaves in woods or under almost any pile of rubbish among decaying vegetable matter. These terrestrial species do, indeed, become rare in the colder parts of the world, but are found as far north as Greenland. Other species less familiar, but perhaps hardly less abundant, inhabit ponds and streams of fresh water, and others are found along the shores of all oceans; yet others abound among the marine vegetation of the shallow waters, or fix themselves upon the bodies, or within the mouths of fishes and other marine animals. Species are found swimming free in the open ocean, and others are brought up from the greatest depths to which the dredge has yet penetrated.

It will be convenient to give here a brief general account of the structure of the animals composing this order, and an explanation of the terms used in their description. Most of our marine species have a greater or less number of the segments at the posterior end of the body coalescent, but in the genus *Cirolana* they are distinct; the animals are, moreover, of large size and very abundant in some localities; reference will therefore be constantly made to the figures of *Cirolana concharum*, on plates IX and X, in illustration of the parts of the animal and of the terms used. A few specimens of this animal will help materially in gaining a knowledge of the structure of the group; or, if specimens of *Cirolana* cannot be obtained, a common "sow-bug" (*Oniscus* or *Porcellio*) may be substituted.

The body appears to consist of fourteen segments, of which the first is the head; the next seven form the thorax, or percion of Spence Bate, and the last six the pleon, sometimes called the abdomen. Returning to the head we find, looking from above, a pair of eyes—each consisting of a group of ocelli—and two pairs of antennary organs. Of these the upper pair, or antennulæ (pl. X, fig. 60), consist on each side of three comparatively large basal segments, which, together, are called the peduncle, or peduncular segments, and support a more slender and tapering flagellum or lash, composed of a considerable number of short segments, decreasing in diameter toward the tip, and each, usually, bearing a fascicle of setæ, which are called by Fritz Müller olfactory setæ, from their supposed function. The antennulæ are very small and rudimentary in "sow-bugs" and their allies. Below the antennulæ are the antennæ properly so called (pl. X, fig. 61 *a*), which are also composed of a peduncle and flagellum. The five basal segments constitute the peduncle, and the following, usually much shorter and smaller segments, are flagellar.

Underneath, the mouth is seen to be protected by a pair of organs called maxillipeds (pl. X, fig. 62 *a*), with which, for convenience of dissection, we shall commence the description of the parts of the mouth. The five terminal segments of the maxillipeds in *Cirolana* (numbered 1 to 5 in the figure) constitute the palpus, but this number varies in the different genera. They are articulated to the external surface of the large basal segment (*m*), usually proportionally much larger than in *Cirolana*, as in *Idotea phosphorea* (pl. V, fig. 28 *b*, *m*), or in the "sow-bug" where the palpus is greatly reduced. The basal segment of the maxilliped is, in general, produced internally beyond the origin of the palpus, and furnished with strongly plumose or pectinated setæ at the tip. Frequently along its inner margin one or more short styliform organs are attached, as in *Jæra albifrons* (pl. I, fig. 5), while along its basal margin is a more or less distinct suture, indicating the epimeral segment of this organ, which will be further explained. The basal segments of the opposite maxillipeds meet along the median line, where their margins are nearly straight, and to the base of the outer margin is attached a more or less triangular external lamella (pl. X, fig. 62 *a*, *l*). The name "maxilliped" is frequently used for the basal segment only, which is often, as in the "sow-bugs," much larger than the rest of the organ and serves to cover and protect the other organs of the mouth.

When the maxillipeds are removed we find two pairs of maxillæ, the outer and inner; of these the outer, or second pair (pl. X, fig. 61 *b*), are in general of a delicate texture, and three-lobed at the tip, the two outer lobes being articulated to the basal piece, and all three lobes ciliated on their inner margins. The inner, or first pair of maxillæ are of a less delicate texture than the outer, and are hardly of the ordinary form in *Cirolana* (pl. X, fig. 61 *c*); reference may, therefore, be made to *Synidotea nodulosa* (pl. VI, fig. 35 *c*), where the two unequal lobes are shown, the inner comparatively small, and supported on a slender peduncle, curved inward, truncated at the tip, and bearing stout, curved, pectinated setæ; the outer much more robust and larger, similar in general outline to the inner, but armed with stout, curved, denticulated spines at the tip.

The mandibles (pl. X, fig. 61 *d*) are usually toothed at the apex, the teeth being supported on a dentigerous lamella, which may be double on one mandible, usually the left, and receive the lamella of the opposite mandible between the two; below this lamella is often a comb of pectinate setæ, and, generally, a molar process, as in *Janira alta* (pl. III, fig. 12 *b*, *m*). In many genera a three-jointed palpus (pl. X, fig. 61 *d*, *p*) is articulated to the external surface of the mandible, and, usually, the terminal segment of the palpus is more or less semicircular, or curved, and bears on its inner margin a very regular comb of setæ (pl. III, fig. 12 *b*), apparently of service in cleansing the organs of the mouth. This comb may be continued or repeated on the second segment, as in *Cirolana* (pl. X, fig. 61 *d*, *p*). In the "sow-bug" and many other genera the mandibles are destitute of palpi. The oral opening between the mandibles is defended by an upper and lower lip, or labrum and labium, which are, however, median, and not paired organs, like the other parts of the mouth.

The seven thoracic segments are of firm texture above, but softer underneath. The dorsal surface is in general more or less rounded, and in *Cirolana* is continued well down at the sides, where, except in the first segment, it is crossed by a suture cutting off a quadrate, or somewhat triangular piece, called an epimeron, or, in the plural, the epimera. The epimera are well shown in the side view of *Cirolana concharum* (pl. IX, fig. 58). They belong to the legs, and form a portion of the large proximal segment called the coxa. Usually, however, the legs are figured as in pl. X, fig. 62 *b*, without this segment, which adheres strongly to the body; often, as in the first segment of *Cirolana*, the suture separating it disappears. The remaining six segments of the legs are more slender, and are called respectively, beginning with the segment following the coxa, the basis, ischium, merus, carpus, propodus, and dactylus, the last being usually slender and curved, often bearing a curved spine or claw at the tip, and, especially in the first pair, capable of flexion on the propodus, so as to form a prehensile hand. In the *Tanaidæ*, as in many of the higher Crustacea, the propodus may be prolonged into a digital process, against which the dactylus closes, forming a chela (pl. XIII, fig. 85), or chelate hand, as in the lobster. In the *Egidæ* and the *Cymothoidæ* a greater or less number of the dactyli are strongly curved or hooked, for the purpose of retaining firm hold of the host, on which these parasitic species live. Legs thus constructed are called ancoral, as in *Liconeca oralis* (pl. XI, fig. 67 *d* and *e*).

Of the seven pairs of legs attached to the thorax or pereion, the first three have in general a resemblance to each other, and are often more or less prehensile, while, as in *Chiridotea* (pl. IV, figs. 16 and 20), the last four are more strictly locomotive organs; but to this condition of things there are many exceptions, especially in the development of the first pair of legs, which are quite variable throughout the order, being not even pediform in the males of the *Gnathiidæ*, but two-jointed, in our species, and lamelliform (pl. XII, fig. 76 *d*). Except in this family, however, no confusion arises from speaking of the thoracic appendages as the first to the seventh pair of legs, or thoracic legs, and in general these terms will be used except where it may be necessary to use the technical terms, gnathopods or gnathopoda and pereiopods or pereiopoda, for these organs, as proposed by Spence Bate, according to whose system the first and second pairs are called the first and second pairs of gnathopoda* or gnathopods, and the remaining five pairs the first to the fifth pair of pereiopoda or pereiopods. When necessary these terms will be added as explanatory, having the merit of scientific accuracy as well as applicability to other groups of Crustacea, where a

---

* See also Edwards, Ann. Sci. nat., III, tome XVI, p. 221–291.

marked distinction of structure and function frequently occurs between the organs homologous with the second and third pairs of legs in the Isopoda.

In the adult females of this order there is commonly formed, on more or less of the under surface of the thorax, an incubatory pouch for the reception and development of the eggs. The outer surface of the pouch is usually formed by four pairs of lamellæ attached just within the origins of the second, third, and fourth, together with the first or fifth pairs of legs, and in the females of many genera, *Sphæroma* and *Asellus* for instance, these lamellæ may be observed in a rudimentary condition on the under surface of the thorax when not actually in use carrying eggs or young. In *Asellus*, and in some other genera, they are found upon the first to the fourth segments, instead of the second to the fifth. In *Anthura* the incubatory pouch extends over only three segments, the third, fourth, and fifth; and in *Astacilla* it is confined to a single segment, being composed of a single pair of elongated plates attached to the fourth segment. In *Tanais* a further remarkable variation occurs, and the eggs and young are carried in sacs attached to the under surface of the fifth thoracic segment, while in the closely allied genus *Leptochelia* the form of the incubatory pouch is normal. In the *Gnathiidæ* and *Anthuridæ*, according to Spence Bate and Dohrn, the incubatory pouch is formed by the splitting of the integument of the inferior surface of the thoracic segments in the females, and for the discharge of the young the outer lamella thus formed further divides into scales, one pair for each segment of the pouch. In *Jæra*, *Epelys*, and probably other genera, a similar mode of development seems to occur.

The six segments of the pleon are smaller than those of the thorax, often much smaller, and frequently more or less united, sometimes consolidated into a single piece with scarcely any trace of division above, but the number of pairs of appendages is generally six, showing the composite nature of the apparently simple organ. Of these six pairs of appendages or pleopods, the first five are more or less concealed beneath the pleon, and consist on each side of a basal segment bearing two lamellæ (pl. IV, fig. 19 c), of which the outer is the anterior when they overlap. These lamellæ, at least the anterior pairs, are usually ciliated along more or less of their distal margins with long slender plumose setæ. In the males of most of the genera, the inner lamella of the second pair bears, articulated near the base of its inner margin, a slender stylet (pl. IV, fig. 19 b, s). This stylet seems to afford, in many cases, specific and even generic characters.

The last segment, sometimes called the telson, has its pair of appendages specially modified, and called the uropods (pl. X, fig. 63). They consist in general like the pleopods of a basal segment bearing two lamellæ, or rami, not being always lamelliform, and in the *Tanaidæ* they are more or less segmented (pl. XIII, fig. 86). One of these rami may disappear, as in *Sphæroma* and in some of the *Idoteidæ* (pl. V, fig. 25 c), where a further modification takes place, and the uropods are so articulated to the inferior surface of the pleon as to fold together like a pair of cupboard doors, forming an operculum for the protection of the more delicate pleopods. Except in the *Tanaidæ*, respiration is carried on by means of the pleopods.

In the *Asellidæ*, *Idoteidæ*, and some other families two or more of the segments of the pleon are united, so that, seen from above, the pleon, like the head, may appear to consist of a single segment, as in *Jæra albifrons* (pl. I, fig. 4), but the number of pairs of its appendages, usually six, remains as evidence of this consolidation. In like manner the head is to be regarded as composed of several segments united, and the number of such segments is indicated by the number of pairs of appendages. In the *Tanaidæ* and many of the higher Crustacea, the eyes, more or less distinctly stalked or articulated with the head, are seen to be of the nature of a pair of appendages, which may be regarded as belonging to the first cephalic segment. The antennulæ and antennæ represent, respectively, the second and third cephalic segments, and, in like manner, the mandibles and two pairs of maxillæ represent the fourth, fifth, and sixth segments of the head. A seventh segment is indicated by the maxillipeds. This segment is regarded by Huxley as properly thoracic[*] instead of cephalic, but, for purposes of description, the segment and its appendages will be regarded as belonging to the head, and the next segment considered the first thoracic.

This segment, like the following thoracic segments, is usually free, and has the dorsal region well developed, but in the adult *Gnathia* it is united with the head, and still more closely so in the *Tanaidæ*. The seventh thoracic segment is the last to develop, and in young Isopoda, taken from the incubatory pouch, only six pairs of legs are commonly found. In *Gnathia* this condition prevails through life, and in the adults the first pair of legs are also modified, especially in the males, so as to quite lose their pediform character, leaving apparently only five pairs of legs. Further modifications of structure will be described in the families and genera in which they occur.

The nomenclature adopted, as explained above, corresponds nearly with that proposed by Mr. C. Spence Bate in his Report on British *Edriophthalma*, and used by the authors of the British Sessile-eyed Crustacea.

The length of an Isopod, in the present article, is given as the length of the body, exclusive of appendages, and is measured from the front of the head to the tip of the pleon. When, as in *Janira*, the head is produced medially into a "rostrum" (see pl. II, figs. 9 and 10), the measurement is taken from the tip of the rostrum, which is a part of the head, and not properly an "appendage."

Among the *Edriophthalma* or sessile-eyed Crustacea, the *Isopoda* may in general be characterized as follows: Body depressed rather than compressed; respiration carried on by means of the pleopods, of which the last pair only are modified into uropods.

---

[*] Huxley, Anat. Inv., Am. ed., p. 276.

The body is said to be depressed, or flattened from above, in distinction from the form usually seen in the *Amphipoda*, where it is in general flattened from side to side. An important exception to the ordinary mode of respiration occurs in the *Tanaidæ*, as has already been mentioned. In this family respiration takes place in two lateral cavities, situated beneath the integument of a large cephalothoracic shield, covering the head and first thoracic segment. In general, as the name of the order indicates, the legs are similar in structure and function throughout, as in the "sow-bug," but may differ considerably, as in the *Arcturidæ*, the *Munnopsidæ*, and the *Tanaidæ*.

The arrangement of the families in the present paper can only be regarded as tentative, and no higher grouping will be attempted further than to indicate briefly the relationships of a few of the families to each other.

The *Oniscidæ* may, on account of their aërial respiration, be regarded as standing quite distinct from the remaining families, and should, perhaps, be further divided as proposed by Kinahan. As they do not, however, come within the proper scope of this article, I have not attempted to subdivide the family. The *Bopyridæ* have been placed near the *Oniscidæ* in deference to the opinions of Dr. Fritz Müller. Having made no study of this family myself I do not express any opinion as to the propriety of separating it so widely from the *Cymothoidæ*, with which it has usually been associated. The *Asellidæ* and *Munnopsidæ* are closely allied to each other. The *Idoteidæ* and *Arcturidæ* form a group distinguished especially by their operculiform uropods. The above families correspond nearly with the "marcheurs" or walking Isopoda of Edwards, and more nearly with the "gehende Asseln" of Müller. They usually have the antennulæ much less developed than the antennæ, and the uropods terminal or inferior, that is, attached to the end of the last segment, or in the last two families to its inferior surface.

The *Sphæromidæ* and *Limnoriidæ* are closely allied, and perhaps ought hardly to be kept separate as families. The *Cirolanidæ*, *Ægidæ*, and *Cymothoidæ* form another group embracing a wide diversity of forms, from the active predatory *Cirolana* to the sedentary and distorted *Livoneca*, and yet apparently connected by easy gradations. The remaining families are generally regarded as aberrant, and form the "Isopoda aberrantia" of Bate and Westwood. They do not present any very evident relationships with the preceding. Of these the *Anthuridæ* have usually been associated with the *Idoteidæ* or the *Arcturidæ*, or with both. Except an elongated form, however, they do not appear to have much in common with either of these families. According to Dohrn's observations they are related to the *Gnathiidæ* in the structure of the incubatory pouch. The *Gnathiidæ* have the head united with the first thoracic segment, as in the *Tanaidæ*, but this last family is widely separated from the others, and doubtless ought to be regarded as forming a distinct suborder, according to the views of Dr. Fritz Müller.

The arrangement of the families adopted, and to a certain extent their affinities, are indicated in the subjoined table, in which, however, as throughout the article, special reference is had to the representatives of the order in New England waters, extralimital species, genera, and even higher groups, *Apseudes* and the Serolids, for example, being disregarded. The arrangement will be seen to considerably resemble that of Dr. Fritz Müller. I have placed the *Tanaidæ* at the other end of the order, partly, however, from the necessity of a lineal arrangement.

SYNOPTICAL TABLE OF FAMILIES.

I. Respiration pleonal; legs not furnished with a chelate hand.
  1. Legs in seven pairs.
    a Antennulæ small or rudimentary; antennæ longer, often much elongated.
      † Uropods terminal, sometimes rudimentary, rami mostly styliform.
        Legs ambulatory; antennulæ rudimentary; respiration aerial.
                                                        I. ONISCIDÆ, p. 305
        Legs prehensile; sexes very unlike; adult forms degenerate; parasitic .... II. BOPYRIDÆ, p. 311
        Legs ambulatory or prehensile; segments of pleon united; antennæ with a multiarticulate flagellum .... III. ASELLIDÆ, p. 312
        Last three pairs of legs natatory; segments of pleon united; antennæ with a multiarticulate flagellum .... IV. MUNNOPSIDÆ, p. 328
      †† Uropods inferior, operculiform.
        Legs prehensile or ambulatory, not ciliated .... V. IDOTEIDÆ, p. 335
        First four pairs of legs ciliated; last three pairs ambulatory.
                                                       VI. ARCTURIDÆ, p. 361
    b Antennulæ and antennæ subequal; body not elongated.
      † Uropods lateral, with one ramus obsolete or subrudimentary.
        Antennulæ and antennæ well developed; pleon of two segments; uropods with one movable ramus .... VII. SPHÆROMIDÆ, p. 367
        Antennulæ and antennæ short; pleon of six segments; outer ramus of uropods small .... VIII. LIMNORIDÆ, p. 374
      †† Uropods lateral, distinctly biramous; rami mostly lamelliform.
        Mouth carnassial; legs not ancoral; antennulæ exposed in front; pleopods ciliated .... IX. CIROLANIDÆ, p. 376
        Mouth suctorial; first three pairs of legs ancoral; antennulæ exposed in front .... X. ÆGIDÆ, p. 382
        Mouth suctorial; legs all ancoral; antennulæ concealed at base by the projecting front; pleopods naked .... XI. CYMOTHOIDÆ, p. 390
    c Antennulæ and antennæ subequal, or antennulæ much the largest in the males; body cylindrical, elongated.
      † Uropods lateral and superior.
        Legs ambulatory or prehensile .... XII. ANTHURIDÆ, p. 396
  2. Legs in the adult in six, apparently only five, pairs.
        Five pairs of legs ambulatory; antennulæ and antennæ subequal.
                                                       XIII. GNATHIDÆ, p. 408
II. Respiration cephalothoracic; first pair of legs terminated by a chelate hand.
        Legs ambulatory and prehensile; head united with the first thoracic segment; antennular flagellum single .... XIV. TANAIDÆ, p. 413

I.—ONISCIDÆ.

Antennulæ rudimentary; legs ambulatory; pleon of six distinct segments, of which the last is small; mandibles without palpi; uropods terminal.*

This large and important group of Isopoda being terrestrial in habit, only a few species are mentioned in this paper. They inhabit moist situations, and are commonly known as "sow-bugs," "pill-bugs," "wood-lice," &c. Several species may often be found under an old board or pile of rubbish. The genus *Ligia* Fabr. inhabits sea-shores, above tide-level, and a few other genera are found under heaps of seaweed, or burrowing in the sand along the shore. Three such species, belonging to as many genera, are here described and figured, but are less fully treated of than the marine species that follow in the other families. Other species, especially of the genus *Porcellio*, may be found in similar situations.

The family may be at once recognized by the apparent possession of only a single pair of antennæ. These are the antennæ properly so called, the antennulæ being minute and rudimentary. This is generally regarded as a character indicating a high degree of development, and causes them to somewhat resemble externally some of the shorter myriopoda, which, like other insects, have but a single pair of antennary organs. The maxillipeds are large and operculiform in this family, with short and few-jointed palpi. The mandibles are destitute of palpi.

The legs are rather weak and fitted only for walking, and usually more or less concealed by the projecting epimeral regions of the thoracic segments. The pleon, in our species, has its segments distinct and decreasing rapidly in size to the last, which bears the more or less exserted uropods. These organs may not, however, project beyond the general outline of the pleon, as they scarcely do in *Actoniscus*, while in *Armadillo* they assist in forming the very regular outline of that part of the body, which closes against the head when those animals, as is their habit, roll themselves into a ball on being alarmed.

This family is placed by Bate and Westwood in a separate "division," the "Æro-spirantia," on account of their aërial respiration. The air, however, requires to be saturated with moisture, and in some of the genera the respiration is, in part at least, aquatic. On this subject the reader is referred to the publications of Duvernoy and Lereboullet and of Nicholas Wagner.

Philoscia Latreille.

Philoscia Latreille, Hist. nat. des Crust. et des Ins., tome vii, p. 45, "1804."

Head rounded in front, not lobed; antennæ with its segments cylindrical, flagellum three-jointed; pleon suddenly narrower than the thorax; uropods exserted, basal segment broad, rami elongate.

---

* The above diagnosis would not include the genera *Tylos* Latreille nor *Helleria* Ebner, which perhaps ought not to be regarded as belonging to this family, although closely allied to it.

This genus may be recognized among our *Oniscidæ* by the rounded head without lobes, and the conspicuously narrowed pleon. Only a single species is as yet known from New England.

Philoscia vittata Say.

> *Philoscia vittata* Say, Jour. Acad. Nat. Sci., vol. i, p. 429, 1818.
> Dekay, Zool. New York, Crust., p. 50, 1844.
> White, List Crust. Brit. Mus., p. 99, 1847.
> Harger, This Report, part i, p. 569 (275), 1874; Proc. U. S. Nat. Mus., 1-79, vol. ii, p. 157, 1879.

PLATE I, FIG. 1.

This species may be recognized, among our terrestrial Isopoda, by the absence of the usual antero-lateral processes on the head, in front of the eyes, and by the sudden contraction of the body at the base of the abdomen or pleon.

Body oval, smooth; about twice as long as broad; head nearly twice as broad as long; eyes large, occupying the antero-lateral regions of the head. The antennulæ are minute and concealed from above. Antennæ minutely hirsute, especially on the last three, or flagellar, segments, inserted below the inner margin of the eyes; first segment short; second about twice as long as the first; third equal in length to the second, clavate; fourth longer cylindrical; fifth longest, slender, cylindrical, straight; flagellum slender, three-jointed, longer than the fifth or last peduncular segment; first flagellar segment about one-half longer than the second; third longer than the second, tapering, tipped with a short transparent filament.

The first thoracic segment is longer than the following ones, which are of about equal length. The anterior angles of the first thoracic segment are somewhat produced at the sides around the head; the posterior angles are broadly rounded. The second and third segments have their posterior angles less broadly rounded, but not at all produced backward. In the fourth segment this angle is scarcely produced, but in the fifth, and still more in the sixth and seventh, it becomes produced and acute. The legs increase in size and length from the first to the seventh pair, and are well armed with spines, especially upon the inferior surfaces of the meral, carpal, and propodal segments. The spines on the latter segment are, however, much smaller than those on the merus and carpus.

The pleon is at the base about two-thirds as wide as the seventh thoracic segment. In the first two segments of the pleon the coxæ, or lateral lamellæ, are short, small, and nearly concealed by the seventh thoracic segment, but in the third, fourth, and fifth segments they are evident and acute but not large. The sixth segment is acute but not prolonged behind, and extends beyond the end of the basal segment of the uropod, which is broad and bears the two rami nearly on the same transverse line. The outer ramus, seen from above, is narrowly and obliquely lanceolate in outline, tapering to the tip, and surpasses by less than half its length the more slender, styliform inner ramus. The uropods, the legs and antennæ, and the segments of the pleon, along their margin, are very minutely hirsute.

The color of these animals is dull and somewhat variable, usually brownish or fuscous, with lighter margins and two broad dorsal vittæ. Length 8ᵐᵐ, breadth 4ᵐᵐ.

This species has been found under rubbish and stones from Great Egg Harbor,! N. J., to Barnstable,! Mass. All the specimens that I have seen have been from the coast, although Say states that it is "very common under stones, wood, &c., in moist situations."

Specimens examined.

Number.	Locality.	Habitat.	When collected.	Received from—	Number of specimens.	Dry. Alc.
1222	Somers and Beesley's Points, N. J.	Shore	—, 1871	A. E. Verrill and S. I. Smith	25	Alc.
1911	Stony Creek, Conn	do		A. E. Verrill		Alc.
2146	Vineyard Sound, Mass	do	—, 1871	U. S. Fish Com	8	Alc.
1919	Barnstable, Mass	do	Aug. 30, 1875	do	3	Alc.

Scyphacella Smith.

Scyphacella, Smith, This Report, part i, p. 567 (273), 1874.

Antenna composed of eight distinct segments, with a geniculation at the articulation of the fourth with the fifth segment; terminal portion, or flagellum, composed of three closely articulated segments besides a minute apical one; mandibles slender; exposed portion of the maxillipeds formed of only two segments.

The genus *Scyphacella* was founded by Professor S. I. Smith, in part I of this Report, for the reception of the following species, the only one yet known. In regard to the relations of the present genus with *Scyphax* Dana[*] Professor Smith says: "This genus differs from *Scyphax* most notably in the form of the maxillipeds, which in *Scyphax* have the terminal segment broad and serrately lobed, while in our genus it is elongated, tapering, and has entire margins. In *Scyphax*, also, the posterior pair of thoracic legs are much smaller than the others, and weak; the last segment of the abdomen is truncated at the apex, and the articulations between the segments of the terminal portion of the antennæ, are much more complete than in our species. The general form and appearance of the genera are the same, and the known species agree remarkably in habits, the *Scyphax*, according to Dana, occurring on the beach of Parua Harbor, New Zealand, and found in the sand by turning it over for the depth of a few inches."

Scyphacella arenicola Smith.

Scyphacella arenicola Smith, This Report, part i, p. 568 (274), 1874.
Verrill, This Report, part i, p. 337 (43), 1874.
Harger, Proc. U. S. Nat. Mus., 1879, vol. ii, p. 157, 1879.

PLATE I, FIG. 2.

The small size, nearly white color, and peculiarly roughened surface of this Isopod will in general serve for its recognition, and the presence

---

[*] U. S. Exploring Expedition, Crustacea, p. 733, pl. 48, fig. 5.

of eyes will further distinguish it from *Platyarthrus*, which is often found inhabiting ants' nests, but would hardly be likely to occur in the sand of the beach.

Body elliptical, pleon not abruptly narrower than the thorax, dorsal surface roughened throughout with small depressed tubercles each giving rise to a minute spinule. Head transverse, not lobed; eyes prominent, round; antennae longer than the breadth of the body; with the first and second segments short; third, fourth, and fifth successively longer and of less diameter; flagellum shorter than the fifth segment, composed of three closely articulated, successively smaller segments, and a very short somewhat spiniform but obtuse terminal one; all the segments, except the minute terminal one, beset with small scattered spinules.

First thoracic segment scarcely embracing the head at the sides; second, third, and fourth segments each about as long as the first, but increasing in breadth; fifth, sixth, and seventh diminishing in length and the last two also in breadth. Posterior lateral angles of the first three segments not at all produced, hardly perceptibly produced in the fourth segment; fifth, sixth, and seventh with the angles increasingly produced but not acute. Legs increasing somewhat in size posteriorly, armed, especially on the inferior surface of the meral, carpal, and propodal segments, with short stout spines.

Segments of the pleon with the coxae but little developed. Terminal segment slightly rounded at the end, not attaining the end of the basal segment of the uropods, which are robust, with the basal segment spinulose, tapering to the base of the short, stout, outer ramus, and bearing the more slender inner ramus much nearer its base. The inner ramus is actually longer than the outer, but being inserted much lower down does not attain the tip of the outer ramus; both are tipped with setae.

"Color, in life, nearly white, with chalky white spots, and scattered, blackish dots arranged irregularly. Eyes black." Length 3.4^{mm}.

This species was "found at Somers and Beesley's Points, on Great Egg Harbor!, New Jersey, in April, 1871, burrowing in the sand of the beaches, just above ordinary high-water mark, in company with several species of *Staphylinidae*," and has also since been found by Professor Smith at Nobska Beach, Vineyard Sound!, Mass., in 1874, and by Mr. V. N. Edwards, on the beach at Nantucket Island!, December 6, 1877. It will doubtless be found at other points along the coast and toward the south.

Specimens examined.

Actoniscus Harger.

Actoniscus Harger, Am. Jour. Sci., III, vol. xv, p. 373, 1878.

Eyes small; antennæ geniculate at the third and fifth segments; flagellum four-jointed; terminal segments of maxillipeds lamelliform, lobed; legs all alike; basal segment of uropods dilated and simulating the coxæ of the preceding segments of the pleon; rami both styliform.

This genus resembles *Actæcia* Dana* MSS., considered as the young of *Scyphax ornatus*, and found with it on the beach at New Zealand. Professor Kinahan,† on the other hand, regarded the genus as indicating a distinct family. The present genus differs from the description and figures of Professor Dana as follows: The flagellum of the antennæ consists of only four distinct segments instead of about six; the terminal segment of the maxillipeds is less distinctly lobed; the inner ramus of the uropods surpasses the outer, instead of falling far short of it; the outer ramus is styliform instead of being enlarged and subequal to the produced and enlarged outer angle of the basal segment.

Actoniscus ellipticus Harger.

Actoniscus ellipticus Harger. Am. Jour. Sci., III, vol. xv, p. 373, 1878; Proc. U. S. Nat. Mus., 1879, vol. ii, p. 157, 1879.

PLATE I, FIG. 3.

This species may be at once recognized by the pleon, which appears to have four pairs of coxæ produced at the sides instead of three, as in *Oniscus* and other genera of this family. The last pair are, however, the basal segments of the caudal stylets, which are of peculiar form in this genus.

The body is oval in outline. The head appears triangular as seen from above, and is angularly produced in a median lobe, but the lateral lobes are also large and divergent, and broadly rounded. The eyes are small, oval, black, and prominent. They are situated at the sides of the median triangular part of the head, and at the base of the lateral lobes. The antennulæ are minute and rudimentary. The antennæ have the basal segment short; the second enlarged distally, especially on the inner side; the third forming an angle with the second, and clavate; the fourth flattened-cylindrical, longer than the third; fifth longest, slender, bent at base and forming an angle with the fourth; flagellum shorter than the last peduncular segment, tipped with setæ and composed of four segments, of which the second and third are equal and longer than the first, while the last is the shortest, and presents indications of another minute rudimentary terminal segment. The maxillipeds have the basal segment nearly twice as long as broad; the terminal segment elongate triangular, ciliated and somewhat lobed near the tip.

---

* U. S. Expl. Exped. Crust., part ii, p. 736, pl. 48, fig. 6 a–h.
† Natural History Review, vol. iv, Proc. Soc., p. 274, 1857.

The first thoracic segment is excavated in front for the head, admitting it about to the eyes. The next five segments are each a little longer than the first, but the last thoracic segment is the shortest. The first segment is dilated at the sides to about twice its length on the median line. The second, and in an increasing degree the succeeding segments are produced backward at the sides. The legs are rather small and weak and of nearly equal size throughout.

The first two segments of the pleon have their lateral processes, or coxæ, obsolete as usual in the family, but the third, fourth, and fifth segments are produced laterally into broad plates, which are close together, and, at their extremities, continue the regular oval outline of the body with scarcely a perceptible break between the thorax and the pleon. This outline is further continued by the expanded basal segments of the uropods, which are even larger than the adjacent coxæ of the fifth segment. At the extremity of the pleon both pairs of rami are visible, the inner springing from near the base of the basal segments below, the outer from a notch near the middle of the inner margin of the basal segment. The rami are tipped with setæ, and the inner just surpass the outer, which, in turn, surpass the produced portion of the basal segments.

Length 4^{mm}, breadth 2^{mm}. Color in life slaty gray.

This species was collected by Professor Verrill, at Savin Rock, near New Haven!, and also at Stony Creek!, Long Island Sound, in company with *Philoscia vittata* Say.

Specimens examined.

Number.	Locality.	Habitat.	When collected.	Received from—	Number of specimens.	Dry. Alc.
2137	Savin Rock, Conn	Shore	—, 1874	A. E. Verrill	2	Alc.
2138	Stony Creek, Conn	do		do	1	Alc.

The genus *Ligia* Fabricius[*] is recorded by Gould[†] from the timbers of a wharf, probably in Boston, and by Dr. Leidy,[‡] with some doubt, from Point Judith, R. I., and the characteristics of the genus are therefore here briefly inserted, as follows:

Antennæ with a multiarticulate flagellum; basal segment of uropods exserted bearing two elongated cylindrical rami.

They are found usually in rocky places and under stones just above high-water mark. They are common on our southern coast, and are probably, at least occasionally, transported by accident within our limits. I have seen no specimens from nearer than Fort Macon, N. C.

---

[*] Suppl. Ent. Syst., p. 296, 1798.
[†] Invert. Mass., p. 337, 1841.
[‡] Jour. Acad. Nat. Sci., II, vol. iii, p. 150, 1855.

II.—BOPYRIDÆ.

This family has not been studied, and only a list of the species, furnished by Professor S. I. Smith, is included. They are parasitic on Crustacea, and at maturity, the females especially, are generally much distorted and degenerate, often losing a great proportion of their appendages. The males are much smaller than the females, and of a more normal form, and they and the young forms must therefore be relied upon to indicate the affinities of this group to the rest of the order. According to Dr. Fritz Müller these forms indicate a relationship to the *Oniscidæ*, and especially to the genus *Ligia*, and in deference to his authority I have inserted them at this place.

Cepon distortus Leidy.

Cepon distortus Leidy, Jour. Acad. Nat. Sci., II, vol. iii, p. 150, pl. xi, figs. 26–32, 1855.
Harger, This Report, part i, p. 573 (279), 1874; Proc. U. S. Nat. Mus., 1879, vol. ii, p. 157, 1879.
Leidya distorta Cornalia and Panceri, Mem. R. Accad. Sci. Torino, II, tom. xix, p. 114, 1861.

"From the branchial cavity of *Gelasimus pugilator*, Atlantic City, New Jersey." (Leidy.)

Gyge Hippolytes Bate and Westwood (Kröyer).

Bopyrus Hippolytes Kröyer, Grönlands Amfipoder, p. 306 (78), pl. iv, fig. 22, 1838; Monog. Fremst. Slægten Hippolyte's nordiske Arter, p. 262, 1842; Voy. en Scand., Crust., pl. xxviii, fig. 2, 1849.
Edwards, Hist. nat. des Crust., iii, p. 283, 1840.
Stimpson, Proc. Acad. Nat. Sci. Philadelphia, 1863, p. 140.
Gyge Hippolytes Bate and Westwood, Brit. Sess. Crust., vol. ii, p. 230, 1868.
Buchholz, Zweite deutsche Nordpolfahrt, p. 286, 1874.
Metzger, Nordseefahrt der Pomm., p. 286, 1875.
Miers, Ann. Mag. Nat. Hist., IV, vol. xx, p. 64, (14), 1877.
Smith in Harger, Proc. U. S. Nat. Mus., 1879, vol. ii, p. 157, 1879.

Massachusetts Bay !, off Salem, on *Hippolyte spinus*, 30 fathoms, sand and mud, August 4, 1877; on *H. Fabricii*, 22 fathoms, gravel, August 4, 1877; on *H. securifrons*, 90 fathoms, soft mud, August 14, 1877. Casco Bay !, on *H. polaris* and *H. pusiola*, 1873. Bay of Fundy !, on *H. spinus* and *H. pusiola*, 1868, 1872. Off Halifax, Nova Scotia, 43 fathoms, September 27, 1877. Gulf of Maine !, 40 miles east of Cape Ann, Massachusetts, on *H. securifrons*, 160 fathoms, soft mud, August 19, 1877; also near Cashe's Ledge, on *H. spina*, 27 and 40 fathoms, rocks and gravel.

East side of Smith's Strait, north latitude 78° 30' (Stimpson). "Discovery Bay," north latitude 81° 44', Greenland (Miers). British Islands (Bate & Westwood). Scandinavian coasts (Kröyer et al.). Spitzbergen (Kröyer).

Phryxus abdominalis Liljeborg (Kröyer).

    Bopyrus abdominalis Kröyer, Nat. Tidsskr., vol. ii, pp. 102, 289, pls. i, ii, 1840;
        Monog. Fremst. Slægten Hippolyte's nordiske Arter, p. 263, 1842; Voy.
        en Scand., Crust., pl. xxix, fig. 1, 1849.
    Phryxus Hippolytes Rathke, Fauna Norwegens, p. 40, pl. ii, figs. 1-10, 1843.
    Phryxus abdominalis Liljeborg, Œfvers. Kongl. Vet.-Akad. Förh., ix, p. 11, 1852.
        Steenstrup and Lütken, Vidensk. Meddelelser, 1861, p. 275 (9).
        Bate and Westwood, Brit. Sessile-eyed Crust., vol. ii, p. 234, 1868.
        Norman, Rep. Brit. Assoc., 1868, p. 288, 1869; Proc. Royal Soc., London,
        vol. xxv, p. 202, 1876.
        Buchholz, Zweite deutsche Nordpolfahrt, p. 287, 1874.
        Metzger, Nordseefahrt der Pomm., p. 286, 1875.
        Miers, Ann. Mag. Nat. Hist., IV, vol. xx, p. 65 (15), 1877.
        Smith in Harger, Proc. U. S. Nat. Mus., 1879, vol. ii, p. 158, 1879.

Massachusetts Bay!, off Salem, on *Pandalus borealis*, *Hippolyte spinus*, and *H. securifrons*, 48–90 fathoms, soft mud, August 13 and 14, 1877; also, on *Pandalus Montagui*, 35 fathoms, mud and clay nodules, August 10, 1877. Cashe's Ledge !, Gulf of Maine, on *Hippolyte pusiola*, 27 and 39 fathoms, rocky, September 5, 1874. Halifax !, Nova Scotia, on *Hippolyte pusiola*, 18 fathoms, fine sand, September 4, 1877; also, on *H. spinus*. About 30 miles south of Halifax !, on *Hippolyte securifrons*, 100 fathoms, fine sand, September 6, 1877.

Grinnell Land, in north latitude 79° 29′; and "Discovery Bay," north latitude 81° 44′ (Miers). Greenland (Kröyer *et al.*). British Islands (Norman *et al.*). Scandinavian coast! (Liljeborg *et al.*). Spitzbergen (Miers).

Dajus Mysidis Kröyer.

    Dajus Mysidis Kröyer, Voy. en Scand., Crust., pl. xxviii, fig. 1, 1849.
        Lütken, Crustacea of Greenland, p. 150, 1875.
        ? G. O. Sars, Arch. Math. Nat., II, ii, p. 354 [254], 1877 ("*D. Mysidis?*").
        Smith in Harger, Proc. U. S. Nat. Mus., 1879, vol. ii, p. 158, 1879.
    Bopyrus Mysidum Packard, Mem. Bost. Soc. Nat. Hist., vol. i, p. 295, pl. viii,
        fig. 5, 1867.
    ? *Leptophryxus Mysidis* Buchholz, Zweite Deutsche Nordpolfahrt, p. 288, pl. ii,
        fig. 2, 1874.

Labrador (Packard). Greenland (Kröyer, Buchholz). ? Off west coast of Norway (G. O. Sars).

Bopyrus, species.

    Bopyrus Leidy, Proc. Acad. Nat. Sci., 1879, pt. ii, p. 198, 1879.
        ? Smith, Trans. Conn. Acad., vol. v, p. 37, 1879.

A species of *Bopyrus* is mentioned by Dr. Leidy as "a parasite of the shrimp, *Palæmonetes vulgaris*," occurring in the summer of 1879, at Atlantic City, N. J.

III.—ASELLIDÆ.

Antennæ elongated with a multiarticulate flagellum; legs ambulatory or prehensile, not strictly natatory; pleon consolidated into a scutiform segment, bearing terminal uropods, which may be nearly obsolete.

This family is represented on our coast by four species belonging to three genera, and a species of another genus (*Asellus communis* Say) is common in the fresh-water ponds and streams of New England. The genus *Limnoria* Leach has been regarded by modern writers as belonging to this family, but will be found in the present article in the *Limnoriidæ* (p. 79). There remain then to be considered the genera *Asellus* Geoffroy,* *Jæra* Leach, *Janira* Leach, and *Munna* Kröyer, which, as represented in our waters, may be further characterized as follows:

The head is well developed, and in *Munna* is of large size; the body is usually depressed or but slightly arched, except that the pleon is vaulted in *Munna*. The eyes are present in our species though not throughout the family. The antennulæ beyond the basal segment are slender and are always much shorter than the antennæ, which are elongated and composed of a five-jointed peduncle and a slender multiarticulate flagellum. The first three peduncular segments are short; the last two elongated. The parts of the mouth are protected below by a pair of maxillipeds with large external lamellæ and five-jointed palpi. Within the maxillipeds are two pairs of maxillæ of the ordinary form; the outer or second pair delicate and three-lobed at the tip; the inner lobe being formed by the projecting basal segment, while the two outer lobes are articulated; all three lobes are provided with curved spiniform setæ. The inner, or first, pair of maxillæ present two narrow lobes; the outer lobe broader and more robust than the inner, and armed with robust curved spines, while the inner is tipped with much weaker setæ. The mandibles (see fig. 12 *b*, pl. III) are provided with one or two acute dentigerous lamellæ (*d*) at the tip, usually a comb of setæ and a strong molar process below (*m*), and a triarticulate palpus (*p*). This latter organ is, however, wanting in the genus *Mancasellus* Harger† from the Great Lakes and other fresh-water localities of North America.

The seven segments of the thorax are distinct from the head and from each other, and differ but little in general appearance throughout. The legs are mostly slender and elongated, except that the first pair may be more robust and better fitted for prehension. In our marine species the dactylus, at least behind the first pair of legs, is short and armed with two small claws or ungues, while the propodus is capable of considerable flexion on the carpus.

The segments of the pleon are united into a single piece, which is scutiform above, flattened or but little arched, except in *Munna*, and bears, at or near the tip, the biramous uropods, which are, however, nearly obsolete in *Munna*. The pleon often shows more or less trace of its compound character in imperfect transverse sutures on the dorsal surface near the base, and below it is excavated for the pleopods, the posterior pairs of which are delicate and branchial in their nature, while the anterior pairs

---

* "Hist. des Ins. t. ii" (Edw.). For information in regard to the common European form of this genus the reader should consult the admirable work of G. O. Sars, Hist. nat. des Crust. d'eau douce de Norvège.

† Am. Jour. Sci., III. vol. xi, p. 304, 1876. See, also, op. cit., vol. vii, p. 601, 1874, and This Report, part ii, p. 659, pl. i, fig. 3, 1874.

are variously modified in the different genera and in the sexes, so that much confusion has been introduced into the family by mistaking sexual for generic modifications of these organs. The branchial pleopods are usually protected by a thickened anterior pair, which, especially in the females of our marine species, may be consolidated into a single opercular plate, as will be further described. The incubatory pouch in the females does not appear to extend farther back than the fourth thoracic segment, and it may be confined to the second, third, and fourth segments.

In the last-mentioned, as well as in many other characters, this family is closely related to the next, and perhaps the *Munnopsidæ* may yet require to be united with it. Our species of the two families are at once distinguished by the last three pairs of legs, which are ambulatory in the *Asellidæ* and natatory in the *Munnopsidæ*. Our *Munnopsidæ* are, moreover, like the other known species of that family, destitute of eyes, while the marine *Asellidæ* have evident or conspicuous eyes, but the fresh-water genus *Cæcidotea* Packard[*] is blind, as are also certain foreign species referred to the present family. The relations of the *Asellidæ* with families other than the *Munnopsidæ* are less evident. They were associated by Professor Dana[†] with his *Armadillidæ* and *Oniscidæ* to form his subtribe *Oniscoidea*, and, *Limnoria* being excluded, the group appears to be a natural one.

Asellus communis Say, confined to fresh waters, and the only known New England representative of the genus, was described and figured by the present author, in Professor S. I. Smith's "Crustacea of the Fresh Waters of the United States," published in part II of this report (page 657, plate I, figure 4). Our marine representatives of the family may be most easily recognized by the consolidated pleon, ambulatory or prehensile legs, none of them natatory, and the slender, elongate antennæ. The genera may be distinguished by means of the following table:

Pleon { flattened above; uropods { short, subrudimentary .... JÆRA, p. 314
                                 { well developed .... JANIRA, p. 319
       { vaulted; head large .... MUNNA, p. 325

Jæra Leach.

Jæra Leach, Ed. Encyc., vol. vii, p. "434" (Am. ed., p. 273), "1813–14."

Antennulæ short, few-jointed; antennæ moderately elongated; mandibles with palpi; first pair of legs similar to the following pairs; lateral margins of the thoracic segments projecting over the bases of the legs; uropods short, rami subrudimentary; pleon protected below in the females by a subcircular plate.

The short uropods and projecting lateral margins of the thoracic segments serve to distinguish this genus from its allies, and other characters of generic importance could doubtless be drawn from the pleon and its appendages, as well as from other parts of the structure, but, as it

---

[*] American Naturalist, vol. v, p. 751, figs. 132, 133, 1871.
[†] Am. Jour. Sci., II, vol. xiv, p. 301, 1852.

is represented in our limits by a single species, I have not been able to separate the generic from the specific characters with confidence, and have therefore described the species without attempting it.

Jæra albifrons Leach.

"*Oniscus albifrons* Montague MSS." (Leach).

Jæra albifrons Leach, Ed. Encyc., vol. vii, p. "434" (Am. ed., p. 273), "1813–14";
  Trans. Linn. Soc., vol. xi, p. 373, 1815.

Samouelle, Ent. Comp., p. 110, 1819.

Desmarest, Dict. Sci. nat., tome xxviii, p. 381, 1823; Consid. Crust., p. 316, 1825.

Latreille, Règne Anim., tome iv, p. 141, 1829.

Edwards, Annot. de Lamarck, tome v, p. 267, 1838; Hist. nat. des Crust., tome iii, p. 150, 1840; Règne Anim., Crust., p. 204, 1849.

Moore, Charlesworth's Mag. Nat. Hist., n. s., vol. iii, p. 294, 1839.

Thompson, Ann. Mag. Nat. Hist., vol. xx, p. 245, 1847.

White, List Crust. Brit. Mus., p. 97, 1847; Brit. Crust. Brit. Mus., p. 69, 1850; Pop. Hist. Brit. Crust., p. 231, 1857.

Lilljeborg, Öfvers. Vet-Akad. Förh., Årg. viii, p. 23; 1851; Ibid., Årg. ix. p. 11, 1852.

Gosse, Man. Mar. Zool., vol. i, p. 136, fig. 243, 1855.

M. Sars, Christ. Vid. Selsk. Forh., 1858, p. 153, 1859.

Bate, Rep. Brit. Assoc., 1860, p. 225, 1861.

G. O. Sars, Reise ved Kyst. af Christ., p. (29), 1866; Christ. Vid. Selsk. Forh., 1871, p. 272, 1872.

Norman, Rep. Brit. Assoc., 1866, p. 197, 1867; ibid, 1868, p. 288, 1869.

Bate and Westwood, Brit. Sess. Crust., vol. ii, p. 317, figure, 1868.

Metzger, J. B. Naturhist. Ges. Hannover, xx, p. 32, 1871; Nordseefahrt der Pomm., 1872–'3, p. 285, 1875.

Parfitt, Trans. Devon. Assoc., 1873, p. (18), "1873."

Stebbing, Jour. Linn. Soc., Zool., vol. xii, p. 149, 1874; Ann. Mag. Nat. Hist., IV, vol. xvii, p. 79, pl. v, figs. 5–6, 1876; Trans. Devon. Assoc., 1879 p. (7), 1879.

Meinert, Crust. Isop. Amph. Dec. Dan., p. 80, "1877." (*Iaira.*)

Harger, Proc. U. S. Nat. Mus., 1879, vol. ii, p. 158, 1879.

Jæra Kröyeri Zaddach, Syn. Crust. Pruss. Prod., p. 11, "1844" (*J. Kröyeri* Edwards ?).

Jæra baltica Fried. Müller, Arch. Naturg., Jahrg. xiv, p. 63, pl. iv, fig. 29, 1848.

Jæra copiosa Stimpson, Mar. Inv. G. Manan, p. 40, pl. iii, fig. 29, 1853.
  Packard, Canad. Nat. and Geol., vol. viii, p. 419, 1863.
  Verrill, Am. Jour. Sci., III, vol. vii, p. 131, 1874; Proc. Amer. Assoc., 1873, p. 360, 1874; This Report, part i, p. 315 (21), 1874.
  Harger, This Report, part i, p. 571 (277), 1874.

Jæra nivalis Packard, Mem. Bost. Soc. Nat. Hist., vol. i, p. 296, 1867. (*J. nivalis* Kröyer ?.)

Asellus Grönlandicus Packard, loc. cit. (*not* of Kröyer).

Jæra marina Möbius, Wirbellos. Thiere der Ostsee, p. 122, 1873; Ann. Mag. Nat. Hist., IV, vol. xii, p. 85, 1873. (*J. marina* Fabricius ?.)

Jæra maculata Parfitt, Trans. Devon. Assoc., 1873, p. "253" (18), "1873."
  Stebbing, Trans. Devon. Assoc., 1879, p. (7) 1879, (*albifrons*).

PLATE I, FIGS. 4–8.

This species is at once distinguished from the other marine Isopoda of our coast by the short uropods, arising from a notch in the end of the subcircular pleon. From the terrestrial forms, which it somewhat resembles, and in company with which it may sometimes be found, the above-mentioned character, joined with the multiarticulate flagellum of the antennae, will serve to distinguish it.

The body is oval and flattened, a little more than twice as long as broad. The head is transverse, broadly excavated on each side over the bases of the antennulae, sparingly ciliated on the lateral margins, with short scattered spine-like unequal cilia or setae, which occur in a similar manner along the entire borders of the animal behind the front margin of the head. The eyes are prominent and black, situated near the posterior margin of the lateral regions of the head. The antennulae are five-jointed, and do not surpass the fourth segment of the antennae; the basal segment is large and separated from its fellow of the opposite side by about twice its diameter; the second segment is about as long as the first, but of much less than half its diameter; third segment shorter than the second, fourth still shorter, fifth tapering, tipped with setae. The first three segments of the antennae are short; the fourth is robust, and about as long as the first three together; the fifth is longest, and is followed by a slender elongated flagellum. The maxillipeds (pl. I, fig. 6) have the external lamella (*l*) short and broad, nearly straight on the inner margin, broadly rounded at the end, and somewhat swollen on the external side; the palpus (*p*) is five-jointed; the first three segments flattened, first short; second dilated internally and ciliated; third ciliate in the inner margin and narrowed to the base of the fourth segment, which is cylindrical; fifth short, conical. The terminal lobe of the maxilliped bears two rows of cilia near the apex, and on the inner side a row of short styliform organs. The outer maxillae (pl. I, fig. 6 *a*) consist of a semioval portion, broad and ciliated at the tip, bearing above the middle two articulated lobes, armed with strong curved setae at the tip. The inner maxillae (pl. I, fig. 6 *b*) are armed with short stout spines, which are strongly spinulose on their inner curved side; inner lobe about half the diameter of the outer. Mandibles with a very much projecting molar process, a comb of pectinated setae, and a dentigerous lamella, or two of them on the left side.

The first three thoracic segments are of about equal length along the median line, and are together nearly equal in length to the last four, which are also subequal along the median line, but the fifth segment appears shorter than the others on account of its short lateral margin, which has both its anterior and posterior angles strongly rounded. The epimeral region of the segments projects at the sides so as to cover the bases of the legs, and is squarish in the first three segments, rounded in the fourth, and still more so in the fifth, and obtusely angulated behind in the sixth and seventh. The legs are similar in form throughout, but increase in length to the last pair. They have the basis rather robust; the ischium shorter and flexed on the basis; the merus subtriangular, and tipped with spines; the carpus and propodus cylindrical, subequal in length, but the carpus of larger diameter than the propodus; the dactylus short, cylindrical, and provided with two terminal hooklets. There are a few scattered spinules and setæ on the segments, especially the merus, carpus, and propodus. In the males the merus and carpus of the sixth and seventh pairs of legs are provided on their inferior margins with close-set slender curved hairs, which extend nearly the whole length of the carpus and over the distal half of the merus.

The pleon is proportionally broader and shorter in the male (pl. I, fig. 8) than in the female (pl. I, fig. 7). It is broadly rounded behind, continuing the outline of the body without break, and is notched at the tip for the insertion of the uropods, which scarcely project beyond the general outline of the body, and consist on each side of a short, stumpy, cylindrical basal segment, a little oblique at the end where it bears two almost rudimentary rami, the inner about twice as large as the outer, and both tipped with a few short setæ. The lateral margin of the pleon, like that of the body generally, is beset with short, scattered, unequal setæ or spinules. Underneath, the pleon is excavated for the branchial pleopods, which are covered and protected below in the females (pl. I, fig. 7) by a large subcircular plate, sparsely minutely ciliated on the margin. In the male (pl. I, fig. 8) the under surface of the pleon presents on each side a small oval plate, with its inner margin overlapped by a median elongated plate, divided by a central suture, which is open distally. This plate is broad at the base, then narrows toward the middle, after which it expands much more rapidly into an outwardly curved and pointed lobe on each side, ciliated at the tip. Between these two lobes the plate is terminated by two transverse, subquadrate and elongated lobes, which are broadest internally where they are separated along the median line. They are excavated on the anterior margin and less so on the posterior margin, sparsely ciliated behind, and conspicuously so with divergent cilia at the outer short, straight margin. In the females the incubatory pouch appears to be confined to the second, third, and fourth segments.

In size as well as coloration this species varies greatly, females being often found with eggs when less than half the size of the specimen figured. They attain a length of 5^{mm} and a breadth of 2^{mm}, but the males are at least one-third smaller and somewhat narrower than the females, the sides being more nearly parallel. In color there is also much variation. A common color is a dark, slaty gray, with dots or small blotches of yellowish, this color prevailing along the anterior margin of the head. Very frequently darker or lighter shades of green occur, and the incubatory pouch of the females is often bright green. Some specimens are very light colored or nearly white, often with two or more transverse dark bands, with considerable contrast in color; others are reddish brown throughout.

I am unable to separate the American form, *Jæra copiosa* Stimpson, from the common English and European species, although they have hitherto been regarded as distinct. I have had no males from any European locality, but through the kindness of the Rev. A. M. Norman I have had an opportunity of comparing females from Oban, Scotland, with our species, and have found no specific differences. The description and figures given by Rev. T. R. R. Stebbing in the Annals and Magazine of Natural History, IV, vol. xvii, p. 79, pl. v, figs. 5 and 6, show a substantial correspondence in the males also, so that I have regarded the species as common to both coasts. Whether the Greenland species *J. nivalis* Kröyer, and the Southern species *J. Kröyeri* Edwards, are also identical with *J. albifrons* or not, I am unable to determine, in the absence of specimens for comparison. M. Sars says that he has seen specimens of *J. albifrons* Leach from Trieste, but regards the Greenland species as distinct. Möbius regards the species as identical from Greenland to the Mediterranean, and unites them under the name *J. marina*. Metzger, following Bate and Westwood, is more conservative, using the name *J. albifrons* Leach. Bate and Westwood regard *J. nivalis* Kröyer and *Oniscus marinus* O. Fabricius as doubtfully identical with *J. albifrons*, and *J. Kröyeri* Edwards as distinct. *J. Kröyeri* Zaddach = *J. baltica* Friedrich Müller appears to be, without doubt, identical with this species, as it is separated by that author from *J. albifrons* Leach only by the position of the eyes, which were incorrectly described by Dr. Leach as close together. I have, therefore, referred these two names to *J. albifrons* as synonyms, as has been done previously by Lilljeborg and others. *J. maculata* Parfitt, a species based almost wholly on color markings, I have referred to *J. albifrons*, following Stebbing, who believes that he is "in accord with the author of the species" in so doing.

This species is common, and in suitable localities abundant, on the whole coast of New England!, and extends as far north as Labrador! at least, where it was collected by Dr. Packard, who regarded it as identical with *J. nivalis* Kröyer. It is found among rocks, algae, and rubbish along the shore, often nearly up to high-water mark, where it may be associated with some of the *Oniscidæ*, to which it has a certain resemblance in form. It occurs "probably" all around the coast of England (Bate and Westwood). I have examined specimens from Oban!, Scotland. It extends to Finmark, on the coast of Norway (M. Sars), and is common on all the coasts of the North Sea (Metzger). It is recorded by Möbius in the Baltic among stones and algae down to a depth of $18\frac{1}{2}$ fathoms. According to M. Sars this species extends to Trieste on the Adriatic, but without specimens I have not attempted to decide in regard to the synonymy of the **Mediterranean species.**

Specimens examined.

Number.	Locality.	Fathoms.	Bottom.	When collected.	Received from—	Specimens. No.	Sex.	Dry. Alc.
1621	New Haven, Conn.			May 1, 1871		20		Alc.
1617	Stony Creek, Conn.					8		Alc.
1616	Noank Harbor, Conn.			—, 1874	U. S. Fish Com.	25		Alc.
1915	Vineyard Sound, Mass.			—, 1871	...do	1	♀	Alc.
1914	...do	L. w.	Under stones	—, 1871	...do	30		Alc.
1920	Provincetown, Mass			—, 1872	...do	50		Alc.
	...do		Shore	—, 1879	...do	60	♂ ♀	Alc.
	...do		...do	Aug. 13, 1879	...do	60	♂ ♀	Alc.
	...do	L. w.		Aug. 13, 1879	...do	15	♀	Alc.
	...do		Eel grass	Aug. 23, 1879	...do	6	♂	Alc.
	Gloucester, Mass		Algæ	—, 1878	...do	30	♀	Alc.
	...do		Tide pools		...do	7	♀ ♂	Alc.
	Casco Bay			—, 1873	...do			Alc.
1919	Eastport, Me.	L. w.	Under stones	1863–1870	A. E. Verrill	7		Alc.
1918	Eastport, Me., Dog Island.		Tide pool	—, 1872	U. S. Fish Com.	5		Alc.
1912	Indian Tickle, Labrador.				A. S. Packard	7		Alc.
519*	Hopedale, Labrador		Stone		...do			Alc.
	Oban, Scotland			—, 1877	Rev. A. M. Norman.	4	♀	Alc.

* *Asellus grönlandicus* Packard, MSS.

Janira Leach.

Janira Leach, Edinb. Encyc., vol. vii, p. "434" (Amer. ed., p. 273), "1813–14".
Asellodes Stimpson, Mar. Inv. Grand Manan, p. 41, 1853.

Body loosely articulated as in *Asellus*; antennulæ slender, with a multiarticulate flagellum; antennæ elongated, with a spine, or scale, on the second segment and with a long multiarticulate flagellum; mandibles palpigerous; lateral margins of the thoracic segments not completely covering the bases of the legs; first pair of legs prehensile; the carpus thickened, and the propodus slender and capable of complete flexion on the carpus; dactylus short and armed with two small ungues, as in the succeeding pairs of legs; uropods well developed, biramous.

This genus is represented on our coast by two species, one of which was originally described by Stimpson under the name *Asellodes alta*. It does not, however, seem to present any generic differences from *Janira maculosa* Leach, the type of the present genus. Stimpson's generic description appears to have been drawn from the male, as he says: "External pair of natatory feet having each two laminæ, like the others, but broader and hardened, so as to perform the office of an operculum." The two inner of these laminæ are, however, united along the median line nearly to the tip, as will be seen below.

Our species of this genus may be further characterized as follows: The body is elongate oval in general outline, between two and three times as long as broad. The eyes are distinct. The head is produced medially into a distinct rostrum, and the antero-lateral angles are also produced, but in the typical species (*J. maculosa* Leach) the head is rounded anteriorly. The basal segment of the antennulæ is enlarged; the second is more slender and cylindrical; the third is short, cylindrical, or slightly clavate, and is followed by a short subglobose segment having the appearance of a fourth peduncular segment. Beyond this, is a slender multiarticulate flagellum, composed of about twenty to thirty segments, the segmentation becoming indistinct toward the base. These segments are provided, except toward the base, with slender "olfactory setæ." The first three segments of the antennæ are short and robust, and the second bears, near its distal end, on the external side above, a triangular scale, or spine, articulated with the segment and directed forward, outward, and somewhat upward; the third segment is comparatively short and small; the fourth and fifth segments are slender and elongated, and the flagellum tapers from the base and is composed of many, 80 to 120 or more, segments. The maxillipeds (see pl. III, fig. 12 a) are broad, with a rhombic-ovate external lamella (l), and a five-jointed palpus (p), of which the first three segments are flattened and expanded internally, where the second and third segments are also ciliated. The last two segments of the palpus are cylindrical, and bent inward toward the median line. The outer maxillæ are rhombic in outline, ciliated and spiny along the inner margin and at the tip, as are also the two slender, curved, articulated lobes. The inner maxillæ consist of the usual curved lobes, armed at the tip with denticulated spines, which are larger, stronger, and more numerous on the outer large lobe. The mandibles are strong, and furnished with an acute dentigerous lamella on the right side, received between two such lamellæ on the left mandible; below is a comb of setæ and a strong molar process. The palpus of the mandible is composed of three subequal segments, the last furnished with a comb of setæ.

The thoracic segments do not greatly exceed the head in transverse diameter, and are subequal, the second, third, and fourth with a lateral emargination. The legs are slender and elongated, ambulatory, or the first pair subprehensile and somewhat shorter than the following pairs. In this pair the carpus is slightly swollen and the propodus is capable of complete flexion upon it. The dactyli are short in all the legs, as compared with the propodi, and capable of only incomplete flexion. They are armed at the tip with two robust unguiform spines.

The pleon is broad and flattened above. The uropods are well developed and consist of a cylindrical or slightly clavate basal segment bearing two rami of which the inner is the larger and longer. The under surface of the pleon is excavated, and in the females is protected beneath by a subcircular operculum, but in the males of *A. alta*, and probably in both species, the thickened opercular plates are three in number, viz, a pair of semi-oval plates at the sides and a more slender median plate presenting traces of a suture along the middle.

In the females, the incubatory pouch is formed of four pairs of plates attached to the coxal segments of the first four pairs of legs. These plates may usually be easily seen when the females are destitute of eggs, being then small, elongate, oval, and lying near the under surface of the thoracic segments.

Janira alta Harger (Stimpson).

 Asellodes alta Stimpson, Mar. Inv. G. Manan, p. 41, pl. iii, fig. 30, 1853.
  Verrill, Am. Jour. Sci., III, vol. vi, p. 439, 1873; vol. vii, pp. 411, 502, 1874; Proc. Amer. Assoc., 1873, p. 350, 1874.
 Janira alta Harger, Proc. U. S. Nat. Mus., 1879, vol. ii, p. 158, 1879.

PLATES II AND III, FIGS. 9, 12, AND 13.

This species may be at once distinguished from the following by the absence of spines in the dorsal and lateral thoracic regions, from all the other known Isopoda of the coast, by the flattened, scutiform and consolidated pleon, bearing well-developed, exerted, biramous uropods, which are, however, fragile. It is more slender than the following species.

The body is elongated oval in outline, nearly three times as long as broad. The head is produced in front into a prominent but short, acute, median spine or rostrum, and the antero-lateral angles are also acutely produced, but are shorter and less acute than the rostrum. The eyes are prominent and black, situated on the upper surface of the head, near the lateral margins. They are elliptical in outline, with the long axes converging toward a point near to, or beyond, the tip of the rostrum. The basal segment of the antennulæ is shorter than the rostrum; the flagellum consists of about thirty segments and does not attain the tip of the fourth antennal segment. The scale on the second segment of the antennæ is short and triangular, does not surpass the following segment, and is tipped with a few slender setæ. The maxillipeds (pl. III, fig. 12 *a*) have the external lamella (*l*) obtusely pointed at the apex and angulated on the outer side, otherwise they resemble the same organs in *J. spinosa*, as do the outer maxillæ, the inner maxillæ, and the mandibles (pl. III, fig. 12 *b*).

The thoracic segments are but little broader than the head, the first three and the last two segments are about equal to each other in length; the fourth and the fifth are somewhat shorter. The lateral margins of the segments do not cover the epimera from above, and none of them are produced at the sides into acute and salient angulations, as in the next species. In the first segment the lateral margins are rounded and the epimera project as an angular tooth on each side in front. In the second, third, and fourth segments the emargination is behind a prominent but narrow lobe at the anterior angle of the segment and the epimera are two-lobed. In the fourth segment the posterior angle is nearly included in the emargination, and in the last three segments the posterior angle is elided and the epimera occupy its place. The legs are elongated and armed with spines, especially on the carpal segments.

The pleon is rounded-hexagonal in outline, minutely and sharply serrate at the sides behind the middle, and undulated over the bases of the uropods on the posterior margin. The uropods are slender, easily detached, and liable to escape observation. They are nearly alike in the two sexes, and consist on each side of an elongate, somewhat curved and clavate basal segment, bearing at the end two rami, of which the inner is nearly as long as the basal segment, the outer somewhat smaller and shorter. The rami are slightly flattened, and, like the basal segment, armed with setæ, especially at the tip. The branchial pleopods are protected in the female by a subcircular operculum (pl. III, fig. 13 *a*). In the male, the inferior surface of the pleon (pl. III, fig. 13 *b*) presents on each side a nearly semicircular plate (*b*), with its inner margin overlapped by a median, elongated, and narrow plate (*c*), marked along the median line by a suture. This plate is broadest near the base, then contracts on each side to beyond the middle, after which it expands slightly. The median suture is open near the tip, and, on each side, is a rounded lobe, separated by a sinus from the produced external angle.

Length of body, exclusive of the antennæ and uropods, 8^{mm}, breadth 3^{mm}. Color in alcohol usually pale or brownish, with small black dots on the upper surface. The under surface is lighter, as are the legs and antennæ, especially toward their distal extremities.

This species is at once distinguished from the common European *J. maculosa* Leach by the form of the head, which is rostrate, and has also the antero-lateral angles strongly salient, while in *J. maculosa* the anterior margin of the head is nearly straight and the angles are not produced. From *Henopomus tricornis* Kröyer,[*] as described and figured by that author, it differs in the elongated uropods.

This species has not been found south of Cape Cod. Dr. Stimpson's specimens were "dredged in soft mud in 40 f. off Long Island, G. M.," in the Bay of Fundy. It was dredged in Massachusetts Bay! in from 54 to 115 fathoms mud, sand, and stones in 1878. In many localities given below in the Gulf of Maine! from 35 to 115 fathoms in 1873, 1874, and 1877, and 120 miles south of Halifax!, N. S., in 120 fathoms gravel and pebbles in 1877. It has also been obtained from several localities in the Bay of Fundy!, in one case at low water on Clark's Ledge, near Eastport, Me. A specimen was collected in 1879, by Mr. Charles Buckley, of the schooner 'H. A. Duncan,' thirty miles east of the Northeast light on Sable Island, adhering to a specimen of *Paragorgia*, from a depth of 160 to 300 fathoms.

---

[*] Naturhist. Tidsskr., II, B. ii, p. 380, 1847; Voy. en Scand., Crust., pl. xxx, figs. 2 *a-q*, "1849."

Specimens examined.

Number.	Locality.	Fathoms.	Bottom.	When collected.	Received from—	Specimens. No.	Sex.	Dry. Alc.
	Gulf of Maine, ESE. from Cape Ann 29-30 miles.	85	Mud, sand, stones	——, 1878	U.S. Fish Com.	1	♀	Alc.
	Gulf of Maine, ESE. from Cape Ann 30-31 miles.	110–115	Mud, stones	——, 1878	...do	1	♀	Alc.
	Gulf of Maine, SE. ½ S. from Cape Ann 6-7 miles.	54–60	Sand, mud	——, 1878	...do	2	♂ ♀	Alc.
1934	Gulf of Maine, SE. from Cape Ann 14 miles.	90	Soft mud	——, 1877	...do	1		Alc.
1923	Gulf of Maine, E. from Cape Ann 140 miles.	112–115	Sand and gravel	——, 1877	...do	1	♂	Alc.
1935	Between Cape Ann and Isles of Shoals	35	Clay, sand, mud	——, 1874	...do	1		Alc.
1924	Gulf of Maine, S. of Cashe's Ledge.	90	Rocky	——, 1873	...do			Alc.
1925	Casco Bay, Me			——, 1873	...do	1		Alc.
	Banquereau			——, 1878	Capt. Collins	1	♀	Alc.
1927	Bay of Fundy, Me			——, 1872	U.S. Fish Com.	3		Alc.
1928	Bay of Fundy, Clark's Ledge.	L. w.-30	Rocky	——, 1872	...do			Alc.
1929	Bay of Fundy, Buckman's Head.			——, 1872	...do			Alc.
1930	Bay of Fundy, off Todd's Head.			Aug. 27, 1872	...do			Alc.
1932	Bay of Fundy, Eastport.			——, 1870	A. E. Verrill	1		Alc.
	Thirty miles east of Northeast light on Sable Island.	100–300	On *Paragorgia*	——, 1879	Mr. C. Buckley	1	♂	Dry.
1933	South of Halifax 120 miles.	190	Gravel and pebbles.	——, 1877	U.S. Fish Com.	1		Alc.

Janira spinosa Harger.

Janira spinosa Harger, Proc. U. S. Nat. Mus., 1879, vol. ii. p. 158, 1879.

This species is well marked among our known Isopoda, by the double row of spines along the back and the acute laciniations or angulations on the lateral margins of the thoracic segments.

The body is robust, the length but little exceeding twice the breadth. The head is broad, and produced in the median line into a prominent acute spine, or rostrum, about as long as the head. The antero-lateral angles are also produced and very acute, but do not extend as far as the rostrum. The eyes are rounded semi-oval, with the long axes converging toward a point near the base of the rostrum. The basal segment of the antennulæ is less than one-third the length of the rostrum. The second segment is about as long as the first, but of only about half its diameter. The flagellum equals, or slightly surpasses, the third antennal segment, and consists of about twelve segments. The scale, or spine, on the second segment of the antennæ is slender and considerably surpasses the third segment. The external lamella of the maxillipeds has the outer angle prominent, though not acute.

The thoracic segments are produced laterally into one or two acute angulations, giving a sharply serrated or dentated outline to the tho racic region. The first segment is shorter than the second; the second, third, and fourth are about equal in length; the fifth is about the length of the first; the sixth and seventh each a little longer. The first segment is acutely produced at the sides, around the sides of the head, and bears, near the middle of the anterior margin, two short spines, situated about half as far apart as are the eyes, and directed upward and somewhat forward. The second segment has both lateral angles produced into triangular acute processes, of which the anterior is more slender than the posterior and directed more strongly forward. The dorsal spines on this segment are a little farther apart and larger than in the first segment. In the third segment the lateral angulations are more nearly equal than in the second segment and directed less strongly forward. In the specimen figured the third segment bears, on the left side, a single broad angulation, apparently representing the posterior, while the anterior is only indicated by a slight irregularity in the outline. Malformations of this kind appear to be common. The dorsal spines on the third segment are much as in the second. On the fourth segment the anterior angulation is longer than the posterior, and both are directed nearly outward. The dorsal spines on the fourth segment are slightly smaller and nearer together than on the third; but, as in all the preceding segments, they are near the anterior border of the segment. The last three segments are acutely produced at the sides into a single angulation, which is directed more and more backward to the last segment. The dorsal spines on the fifth segment are situated nearer together than on the anterior segments, and rather behind the middle of the segment; they are also smaller than on the preceding segments. On the last two segments they are near the posterior border of the segment, and become somewhat smaller and nearer together on the last segment. The legs are armed with but few, and rather weak, spines.

The pleon is broadest near the base and tapers posteriorly, where the angles are acutely produced; between these angles the margin is rounded and arched over the bases of the uropods, which are about as long as the pleon and less spiny than in *J. alta*. The lateral margin of the pleon is armed with very minute acute spinules, and under a higher power the margins of the thoracic segments and of the head are seen to be similarly armed, especially where most exposed.

Length 8^{mm}, breadth 3.8^{mm}; color in alcohol, white.

This species is near *Janira laciniata* G. O. Sars,[*] but is distinguished by the double row of dorsal spines, whereas Sars says of that species, "Superficies dorsalis medio leviter convexa spinis singulis tenuibus ornata."

The only specimens yet known are two females, which were taken adhering to the cable of the schooner 'Marion', by Captain J. W. Collins, at Banquereau, August 25, 1878.

---

[*] Chr. Vid. Selsk., 1872, p. 92, 1873.

Munna Kröyer.

Munna Kröyer, Naturhist. Tidsskr., B. ii, p. 615, 1839.

Form of the female dilated oval, of the male elongated sublinear; head very broad (about twice as broad as long), in length equal to one-fourth or one-fifth the length of the animal; eyes occupying the postero-lateral angles of the head, prominent, as if pedunculated but not movable; antennulæ inserted above the antennæ and partly covering their bases, short, a little longer than the head, with a four-jointed peduncle and a few-jointed flagellum; antennæ elongated, equaling or surpassing the length of the body, with a multiarticulate flagellum; mandibles with a three-jointed palpus; maxillipeds with a five-jointed palpus; legs all armed with two terminal ungues; first pair shorter and more robust than the others, with a prehensile hand formed of the propodus and the dactylus; the remaining pairs ambulatory, increasing gradually in length, so that the last pair equal or surpass the body in length. The segments of the pleon are united into a single vaulted segment, and its inferior surface is covered, in the females, by a single opercular plate, while in the males the operculum is composed of three parts, as in the preceding genera.

The generic description as given above is in part taken from Kröyer, the author of the genus. The specimens hitherto obtained do not appear to be separable from his species *M. Fabricii*, to which I have therefore referred them, although differing somewhat from each other. The material has unfortunately been, most of it, in poor condition, many of the specimens having been dried and much broken.

Munna Fabricii Kröyer.

Munna Fabricii Kröyer, Nat. Hist. Tidsskr., II, B. ii, p. 380, 1847; Voy. en Scand., Crust., pl. xxxi, figs. 1 *a-q*. "1-49".
Reinhardt, Grönlands Krebsdyr., p. 35, 1857.
M. Sars, Christ. Vid. Selsk. Forh., 1858, p. 154, 1859.
Lütken, Greenland Crust., p. 150, 1875.
Harger, Proc. U. S. Nat. Mus., 1879, vol. ii, p. 159, 1879.
Munna, species, Verrill, Am. Jour. Sci., III, vol. vii, p. 133, 1874; Proc. Am. Assoc., 1873, p. 371, 1874.
? *Munna Bœckii* G. O. Sars, Arch. Math. Nat., B. ii, p. 353 [253], 1877. (*M. Bœckii* Kröyer?)

PLATE III. FIG. 14.

This species may be at once distinguished from anything else known on our coast by the prominent, as if pedunculated, but immovable, eyes, on the posterior lateral angles of the large head, together with the elongated and slender ambulatory legs in seven pairs, the first pair only being somewhat shorter.

The first specimens obtained in a recognizable condition were small and differed somewhat from later specimens, especially in size and proportions; the differences, however, do not appear to be necessarily other than what might be due to age and size, and are such as are described by Kröyer in his specimens of *M. Fabricii*. The legs in the small specimen figured are considerably shorter than in larger specimens obtained in 1878, and the flagellum of the antennulæ consists in the small specimens of a single segment, or with traces of subdivision into two, while in the large specimens it is four-jointed, with a rudimentary terminal segment.

The body is in the female elongate oval, tapering posteriorly, and broadest at the third thoracic segment, where the breadth is equal to about half the length. The males are more slender, and are not dilated behind the head. The head forms about one-fifth of the total length, and is nearly twice as broad as long. Its anterior portion between the bases of the antennulæ and antennæ is comparatively narrow on its upper surface, and is rounded or obtusely angled in front. Behind the bases of the antennulæ it is suddenly much dilated at the sides, and a little behind the dilation are the prominent, strongly convex and laterally projecting eyes, immediately behind which the head contracts suddenly in width, and is then slightly rounded behind. The antennulæ arise in a deep sinus on the antero-lateral region of the head. They consist of a four-jointed peduncle followed by a four-jointed flagellum of about the same length as the peduncle. The basal antennular segment is stout, and subtrigonal in form; the second is more slender and cylindrical, while the third and fourth are subequal, quite short and small, together not over half as long as the second segment, and should perhaps rather be regarded as flagellar segments. The four flagellar segments are of a little less diameter than the last two peduncular segments, and are long and cylindrical, the fourth being tipped with a rudimentary segment bearing two strong terminal setæ. The antennæ are much larger and stouter than the antennulæ and are about two or three times as long as the body. They are composed of a five-jointed peduncle and a slender multiarticulate flagellum. They arise nearly in front of the antennulæ and their first three segments are short and stout, not longer taken together than the first two antennular segments. The fourth segment of the antennæ is only about half the diameter of the first three segments, but is greatly elongated, nearly or quite equaling in length the head and thorax taken together, and is cylindrical, and provided with a few short setæ, especially at the tip. The fifth, or last peduncular, segment is slightly more slender and elongated than the fourth, and is followed by a slender tapering flagellum composed of about seventy-five segments, or, perhaps, in perfect specimens, of a greater number. The maxillipeds are large and broad, as required by the large head, and are furnished with a five-jointed palpus, with the basal segment short, the second and third flattened and expanded internally, where they are also ciliated; the fourth narrow; the fifth short, and both provided with scattered setæ, especially toward the tip.

The first thoracic segment is a little shorter than the second, which is about equal in length to the third and the fourth; the last three segments progressively decrease in length and width, and the seventh is somewhat concealed at the sides by the swollen base of the pleon. The basal segments of all the legs are much alike in form, and differ but little in size throughout. They are cylindrical or slightly clavate, the first pair perceptibly shorter and smaller than the second, from which they increase very slightly to the sixth, which is the largest, the seventh not being larger than the second. The legs disarticulate easily at the end of the basal segment, and in the specimens examined nearly all are broken off at this point. Beyond the basal segment the first pair are comparatively short, about half the length of the body. The ischium of the first pair is robust, and a little longer than the merus; the carpus is subtriangular and armed with strong short spines on its palmar margin; the propodus is about as long as the ischium, slightly swollen, and armed with a few spines; the dactylus is short and armed at the end with two stout curved claws, of which the outer is about twice the length of the inner; between the claws is a slender bristle. The second and following pairs of legs are much more elongated than the first pair, the elongation being principally in the carpus and propodus, and, in a less degree, in the ischium and merus, while the dactylus is comparatively but little elongated. In the second pair of legs the propodus is not longer than the carpus, but it becomes proportionally, as well as absolutely, longer in the following pairs until, in the sixth pair, it may be nearly or quite as long as the body and form about two-fifths the whole length of the leg. The dactyli are, in all the legs, comparatively short, often less than one-tenth the length of the propodus, and armed with two unequal claws, of which the longer is about two-thirds as long as the dactylus itself, and the shorter is more than half the length of the longer. In all the legs the ischium is armed with a few short curved spinules, and the elongated propodal segments are furnished with scattered, slender and elongated, straight spines, each with a minute bristle near the apex.

The pleon is remarkably swollen near the base, and is somewhat pear-shaped; posteriorly it is deep, and bears the uniarticulate uropods in shallow grooves near the end. On the upper surface are a few straight slender spines, and below it is covered in the females by an ovate, obtusely-pointed opercular plate, and in the males by a trifid operculum, the median portion being slender, with nearly parallel sides and a central suture, and the two lateral portions slender, semi-ovate and pointed behind. The pleon appears to be carried habitually, during life, flexed upward at a considerable angle.

The length of the specimen figured, by Mr. Emerton (pl. III, fig. 14), is 1.2^{mm}, breadth 0.7^{mm}; but specimens obtained in 1878 measure 3.1^{mm} in length, 1.5^{mm} in width, in the female, and 1.1^{mm} in the male. The pleon measures in length 1.1^{mm} and in width 0.8^{mm} in the larger individuals.

A single much mutilated specimen of this species was dredged in 12 fathoms, South Bay, Eastport!, in 1872, by the United States Fish Commission, and two more specimens, both females, were obtained on eelgrass in Casco Bay! in 1873. Five specimens were obtained adhering to dried specimens of *Acanella* from 150 fathoms, Western Bank!, in 1878, and a sixth, in 53 fathoms, on Brown's Bank!, in lat. 42° 50′ N., lon. 65° 10′ E., by Captain J. Q. Getchell, of the schooner 'Otis P. Lord,' in the same year. In 1879 a specimen was obtained adhering to *Acanthogorgia armata*, by Captain George A. Johnson and crew of the schooner 'Augusta H. Johnson,' on Western Bank!, in lat. 43° 15′ N., lon. 50° 20′ E., 200 fathoms. These specimens were, as has been mentioned, considerably larger than those at first obtained. Kröyer's specimens were from a depth of 50 fathoms, at Godthaab, Southern Greenland, and according to M. Sars the species is abundant on the coast of Finmark among Hydroids in the coralline zone. G. O. Sars records *M. Bœckii* Kröyer, which he regards as scarcely differing from this species, at the harbor of Reikjavik, Iceland.

Specimens examined.

Number.	Locality.	Fathoms.	Bottom.	When collected.	Received from—	Specimens.		Dry. Alc.
						No.	Sex.	
2141	Casco Bay, Me		Eel-grass	—, 1873	U. S. Fish Com.	2	♀	Alc.
1936	Bay of Fundy, Me	12		—, 1872	... do	1	♀	Alc.
	Brown's Bank	53		—, 1878	Capt. J. Q. Getchell.	1	♀	Dry.
	Western Bank	150	On *Acanella*	—, 1878		5	♂♀	Dry.
	Western Bank	200	On *Acanthogorgia armata*.	—, 1879	Capt. G. A. Johnson.	1	♀	Dry.

IV.—MUNNOPSIDÆ.

In this family the body consists of two more or less distinct divisions, the first consisting of the head and anterior four thoracic segments, and the second of the last three thoracic segments, and the pleon, which is consolidated into a single segment, convex above. The eyes are wanting. The antennulæ are much shorter and smaller than the antennæ, and have their basal segment lamelliform. The antennæ are much elongated, with a five-jointed peduncle, of which the first three segments are short and the last two elongated and tipped with a long multi-articulate flagellum. The maxillipeds have their basal segments flattened and operculiform, covering the other mouth parts, and furnished with a large external lamella and a five-jointed palpus. The first pair of legs are shorter than the three following pairs and imperfectly prehensile. The next three pairs are ambulatory and usually greatly elongated. The last three pairs of legs, or at least the fifth and sixth pairs, are different in form from the preceding, and fitted for swimming, with some of the distal segments flattened and provided with marginal cilia or spines. The pleopods are protected by a thickened opercular plate, and the uropods are short and simple or biramous. The incubatory pouch in the females is beneath the first four thoracic segments.

Of this family, two species have been found on the New England coast, and a third, from the Gulf of St. Lawrence, is here included. The specimens obtained have been mostly in poor condition, and one of these, belonging apparently to an undescribed species, is so imperfect that I have decided to await the collection of better specimens before attempting a specific description. In the family characters given above, as well as in the following generic and specific descriptions, I have availed myself largely of the admirable works of M. Sars and his son G. O. Sars, the distinguished Norwegian naturalists, to whom science is indebted for the discovery and characterization of the present group.

The *Munnopsidæ* of our coast may be easily recognized as belonging to the family by the structure of the last three pairs of thoracic legs, which are fitted for swimming by being more or less flattened and ciliated; the last pair, however, may return to the more normal type of leg, so that the fifth and sixth pairs only may be natatory. The three genera which appear to be represented are distinguished as follows: Body suddenly constricted and slender behind the fourth thoracic segment in *Munnopsis* (p. 329); pretty regularly oval in form, with three pairs of flattened natatory legs in *Eurycope* (p 38); suboval but deeply incised behind the fourth segment, in *Ilyarachna* (p. 40), in which genus the last pair of legs are scarcely at all flattened or ciliated.

Munnopsis M. Sars.

Munnopsis M. Sars, Christ. Vid. Selsk. Forh., 1860, p. 84, 1861; Christ-fjord Fauna, p. 70, 1868.

Anterior division of the body dilated, posterior suddenly much narrower and linear. Antennulæ with the basal segment large and flattened, the flagellum elongate and multiarticulate; antennæ very long and slender, many times longer than the body; the last two peduncular segments greatly elongated; the flagellum about equal in length to the peduncle; mandibles subtriangular, entire and acuminate at the apex, without a molar process; the palpus slender with the last segment thick at the base and curved in the form of a hook; penultimate segment of the maxilliped not dilated inwardly; last segment very narrow and linear. Four anterior thoracic segments excavated above, obtusely rounded at the sides; the three following subcylindrical with short acuminate lateral processes; first four pairs of thoracic legs six-jointed (beyond the coxal segment), the first pair short; the second pair not much longer, rather robust and subprehensile in the males; the two following pairs greatly elongated and very slender, many times longer than the body; but with the basis, ischium, and merus very short; last three pairs of legs natatory, all alike, six-jointed, being destitute of dactyli, with the last two segments, the carpus and propodus, foliaceous, margined with long, slender, delicately plumose setæ. Pleon elongate, much longer than broad; abdominal operculum large (nearly covering the whole under surface of the pleon), suboval, simple in the female, but consisting of three distinct segments in the male, one median and very slender, and two lateral, and furnished within with a peculiar curved organ, terminated behind with a much elongated seta; uropods slender uniramous.

Munnopsis typica M. Sars.

Munnopsis typica M. Sars, Chr. Vid. Selsk. Forh., 1860, p. 84, 1861; Christ. Fjord. Fauna, p. (70), pl. vi–vii, figs. 101–138, 1868; Chr. Vid. Selsk. Forh., 1868, p. 261, 1869.

G. O. Sars, Chr. Vid. Selsk. Forh., 1863, p. 206, 1864; Reise ved Kyst. af Christ., p. (5), 1866; Christ. Fjord Dybvands-fauna, p. (44), 1869; Chr. Vid. Selsk. Forh., 1872, p. 79, 1873; Arch. Math. Nat., B. ii, p. 353 [253], 1877.

Whiteaves, Ann. Mag. Nat. Hist., IV, vol. x, p. 347, 1872; Deep-sea Dredging, Gulf of St. Lawrence (1872), pp. 6, 15, 1873; Am. Jour. Sci., III, vol. vii, p. 213, 1874; Further Deep-sea Dredging, Gulf of St. Lawrence (1873), p. 15, 1874.

Buchholz, Zweite Deutsche Nordpolfahrt, Crust., p. 285, 1874.

Heller, Denksch. Acad. Wiss. Wien, B. xxxv, p. (14) 38, 1875.

Norman, Proc. Royal Soc., vol. xxv, p. 208, 1876.

Miers, Ann. Mag. Nat. Hist., IV, vol. xx, p. 65, 1877.

Harger, Proc. U. S. Nat. Mus., 1879, vol. ii, p. 159, 1879.

PLATE II, FIG. 11.

This species is easily recognized among the known Isopoda of our coast by the form of the body, which suddenly diminishes in diameter behind the fourth thoracic segment, so that the last three thoracic segments, bearing the ciliated, swimming legs, are only about half as broad as the anterior part of the body.

Anterior division of the body depressed, posterior subcylindrical; breadth of body less than half the length. Head small, with the length and breadth about equal, equaling the two anterior thoracic segments in length, but of much less breadth, truncate in front and without a rostrum, bearing near the posterior dorsal margin two minute conical tubercles. The eyes are wanting. The antennulæ in the female, when reflexed, extend to the third thoracic segment, in the male to the fourth, with the flagellum longer than the peduncle, pectinate or furnished with a longitudinal series of long setæ, multiarticulate; segments in the female, 23 to 28; in the male, 65 to 66. The antennæ are greatly elongate, about five times as long as the body, very slender; peduncle more than twice the length of the body, the last two peduncular segments beset with numerous short spinules, arranged in longitudinal rows; flagellum nearly as long as the peduncle, composed of about 130 segments. The external lamella (*l*) of the maxillipeds (pl. II, fig. 11 *b*) is narrowed in front with the external margin convex.

The four anterior thoracic segments are subequal, short, about five times broader than long; last three segments broader than long, less than half the width of the preceding segments, bearing near the anterior dorsal margin two small conical tubercles; pleon slightly longer than the three preceding segments together, but not narrower, forming somewhat more than one-fourth the length of the body, elongate-suboval, the breadth scarcely equaling half the length, with a median, rounded, dorsal crest, but little elevated, and bearing in front of this near the anterior margin a small conical tubercle.

Propodus shorter than the carpus in the first pair of legs, equal to it in length in the second pair, which in the males (pl. II, fig. 11 e) have the carpus thickened, and armed, on the inferior margin, with stronger spines than in the females; third and fourth pairs of legs about thrice the length of the body, with the three basal segments, basis, ischium, and merus, very short and robust; the last three very much elongated and filiform; the propodus longer than the carpus, both armed with many short spinules arranged longitudinally; dactylus about one-fifth as long as the propodus, slightly curved, naked, very minutely serrulate along the convex margin. Last three legs (pl. II, fig. 11 f) with the carpus and propodus elongate-subelliptic, both segments strongly ciliated, the propodus a little shorter than the carpus.

Abdominal operculum in the female (pl. II, fig. 11 g) with a longitudinal, elevated, acute median crest, flattened medially in the males. Uropods slightly more than one-third the length of the pleon, composed of two subequal segments. Laminæ of the incubatory pouch in the females attached to the anterior four thoracic segments; the three posterior pairs large; the third and fourth suborbicular; the second elongate; the first much smaller, bifid at the apex.

Length 8–10^{mm}; antennæ 40–50^{mm}; third and fourth pairs of legs 24–30^{mm}. Color, light yellowish, or grayish, in alcohol; lighter below.

The specimens that I have had an opportunity of examining were all more or less imperfect, and I have therefore, in both the generic and specific descriptions given above, made free use of the admirable and exhaustive description of this genus and species by M. Sars,[*] and the figures of the species on plate II were copied from the same author, having been drawn by his not less distinguished son, G. O. Sars.

This species like its allies is an inhabitant of deep water on muddy bottoms. Three specimens, the only ones that I have personally examined, were taken by the Fish Commission in the Bay of Fundy! between Head Harbor and the Wolves, in 60 fathoms muddy bottom, August 16, 1872. It has been dredged by Mr. Whiteaves in the Gulf of St. Lawrence in 125 to 220 fathoms; by the Valorous Expedition in Baffin Bay in 100 fathoms (Norman); in 25 to 50 fathoms off Cape Napoleon, Grinnell Land, by the Arctic Expedition (Miers); between Norway and Iceland in from 220 to 417 fathoms; Christiania fiord, 200 to 230 fathoms (G. O. Sars); Christiania Sound 50 to 60 fathoms,

---

[*] Bidrag til Kundskab om Christiania-fjordens Fauna, 1868, pp. 70–95, pls. vi–vii. (Nyt Magazin.)

whence the species was described by M. Sars; off Storeggen, 400 fathoms (G. O. Sars), and northward among the Loffoden Islands, 250 fathoms; the coast of Finmark, Spitzbergen (Buchholz), and the Arctic Ocean about Nova Zembla (G. O. Sars.)

Eurycope G. O. Sars.

Eurycope G. O. Sars, Chr. Vid. Selsk. Forh., 1863, p. 208, 1864.

Body depressed, subovate as seen from above; about equally attenuated before and behind. Head of medium size, more or less produced between the antennulæ; antennæ very slender, two to four times as long as the body; flagellum longer than the peduncle; mandibles robust, quadridentate at the apex, and bearing below a series of rigid setæ and a strong molar process; mandibular palpus well developed, with the terminal segment enlarged at its base and curved. Four anterior thoracic segments subequal, short; three posterior segments large not suddenly narrower than the anterior segments; the first pair of legs shorter than the next three, with the dactylus short; the next three pairs elongated, and with elongated and slender dactyli; three posterior pairs of legs distinctly natatory, with the carpus and propodus strongly flattened and provided with numerous plumose marginal setæ; dactylus of the ordinary form. Pleon rather large, broader than long, obtusely rounded behind; operculum subpentagonal with rounded angles, much smaller than the pleon. Uropods short, biramous, rami uniarticulate. Dorsal surface of the body smooth and shining.

For the characterization of the genus, as given above, I have depended largely upon the work of G. O. Sars, having had myself, for examination, only the following species:

Eurycope robusta Harger.

Eurycope robusta Harger, Am. Jour. Sci., III, vol. xv, p. 375, 1878; Proc. U. S. Nat. Mus., 1879, vol. ii, p. 159, 1879.

PLATE III, FIG. 15.

This species may be recognized by the flattened and ciliated swimming legs, in three pairs, on the last three thoracic segments, which are not, as in the preceding species, suddenly of much less diameter than the anterior four segments.

Body oval with the length equal to, or slightly exceeding, twice the breadth. Head, behind the bases of the antennulæ, longer than the first thoracic segment, produced medially into a short rostrum about half as long as the basal antennular segment. Antennulæ (pl. III, fig. 15 *a*) attaining the middle of the fourth segment of the antennæ in the females, surpassing the middle of this segment in the males; basal segment subquadrate, spinulose at the distal angles, somewhat narrowed from the base, bearing the second much smaller segment a little beyond the middle of its superior surface; third segment longer and more slender than the second; flagellum of more than twenty articulations, which become indistinct near the base, and are furnished with terminal setæ. Antennæ about thrice the length of the body in the female, somewhat shorter in the male, the sexes differing in the fourth and fifth segments, which, in the females, are subequal in length and, together, as long as the body, while in the male the fifth is shorter than the fourth, and the two segments together are about two-thirds as long as the body. The flagellum is long, slender, and multiarticulate. Maxillipeds (pl. III, fig. 15 b) with the external lamella sub-rhombic, emarginate on the exterior distal side; palpus five-jointed, first segment short, produced externally into a very acute angle; second and third segments broad and flattened; fourth narrow with the inner angle produced and rounded; fifth short, oval. Maxillæ of the ordinary form, outer pair with slender lobes. Mandibular palpus elongated, last segment strongly curved.

Thorax widest at the fourth segment; first four segments forming about one-third its length on the median line, last segment longest, all with their antero-lateral angles produced, the anterior four with the epimera projecting as an acute process below, and in front of, the angle. First pair of legs (pl. III, fig. 15 d and d') about three-fourths the length of the body; dactylus short; propodus shorter than the carpus; slightly hairy, especially on the propodus with slender hairs. Next three pairs of legs longer than the body, subequal, but increasing a little in length to the fourth; dactyli slender and acicular; propodi and carpi subequal, spinulose along their inner margins in the second pair, but not in the third and fourth. Last three pairs of legs with the carpus strongly dilated and flattened, subcircular as seen in pl. III, fig. 15 f, where the sixth pair is represented; propodus also much flattened and dilated; both segments strongly ciliated with plumose bristles, as is also the ischium, or second segment along the outer dilated margin; dactylus about half the length of the propodus instead of less than one-third its length, as in *E. cornuta* G. O. Sars, the species most resembling the present.

Pleon much broader than long, broadly rounded behind. Operculum also broader than long, strongly roof-shaped. Uropods (pl. III, fig. 15 g) with the basal segment shorter than the rami, which are uniarticulate, cylindrical, of equal length, obtuse and tipped with a coronet of short spines. The inner ramus is more robust, but not longer than, the outer.

Color in alcohol, honey yellow; length 4.5^{mm}; breadth 2.2^{mm}.

This species appears to approach *E. cornuta* G. O. Sars,* but may be readily distinguished by its greater size, by the shortness of the rostrum, the equal rami of the uropods, and the shape of the external lamella of the maxillipeds, which he describes in that species as "versus apicem dilatata et emarginata utrinque acute producta." In the third and fourth pairs of legs, moreover, the carpus and propodus are not armed with spines as in that species according to Sars' description.

---

* Chr. Vid. Selsk. Forh., 1863, p. 209, 1-64.

This species was dredged by Mr. J. F. Whiteaves in the Gulf of St. Lawrence! at a depth of 220 fathoms muddy bottom, and has not yet been found on the coast of New England. It is introduced here from the probability that it will yet be discovered in the deeper parts of the Bay of Fundy, where the allied *Munnopsis typica* M. Sars has already been found, or even in the Gulf of Maine.

Specimens examined.

Number.	Locality.	Fathoms.	Bottom.	When collected.	Received from—	Specimens		Dry. Alc.
						No.	Sex.	
1938	Gulf of St. Lawrence	220	Mud		J.F.Whiteaves	10	♂ ♀	Alc.
1939	Gulf of St. Lawrence	220	....do		....do	3		Alc.

Ilyarachna G. O. Sars.

Mesostenus G. O. Sars, Chr. Vid. Selsk. Forh., 1863, p. 211, 1864.
Ilyarachna G. O. Sars, Christ. Fjord. Dybvands-fauna, p. (44), 1869.

Body scarcely depressed, subpyriform as seen from above, narrowed behind; its anterior division separated from the posterior by a deep constriction. The head is large and broad and without a rostrum. Antennulæ short, with a flagellum composed of but few segments. Antennæ exceeding the body in length, with a multiarticulate flagellum. Mandibles short and strong, entire at the apex; molar process armed with a few setiform spines; palpus either small and three-jointed or wanting. Four anterior thoracic segments short, excavated above and furnished with lateral processes directed forward; the three following convex above and destitute of lateral processes; the antepenultimate scarcely narrower than the anterior segments and deeply emarginate behind. First pair of legs nearly as in the preceding genus; second pair unlike the others and usually more robust; the following two subequal and commonly much elongated; fifth and sixth pairs of legs much as in *Eurycope;* the last pair unlike the preceding, long and slender, with the segments scarcely flattened, and armed with a long curved claw. Pleon narrowly triangular, pointed at the apex. Abdominal operculum large, covering nearly the whole of the under surface of the pleon, provided with a median crest and numerous marginal setæ. Uropods simple, appressed to the pleon.

For the generic description given above I have depended almost entirely upon the work of Dr. G. O. Sars, who originally described the genus under the name *Mesostenus*. That name being preoccupied he subsequently changed it to *Ilyarachna*.

Ilyarachna species.

A single imperfect specimen of a species apparently belonging to this genus was dredged in 106 fathoms, gray mud, 21 miles east of Cape Cod Light!, September 18, 1879. The species is probably yet undescribed, but, in view of the very imperfect condition of the only specimen yet known, I have decided to await the collection of better specimens before attempting to make out its characters. It may yet be found to represent an undescribed genus, but I am at present inclined to regard it as a species of *Ilyarachna*.

V.—IDOTEIDÆ.

Antennulæ consisting of four segments, of which the basal is more or less enlarged and the terminal clavate; mandibles not palpigerous; thoracic segments subequal in length; pleon with more or fewer of its segments consolidated into a large, scutiform, terminal piece; uropods inferior, transformed into a two-valved operculum protecting the pleopods.

The *Idoteidæ* are represented on the New England coast by ten species; another, found near our northern limits, is included, making eleven in all, belonging to five genera. The family may be further characterized, so far as regards our species, as follows: The body is depressed, and varies in its proportions of length to breadth from about two to one in *Chiridotea cæca* to nearly six to one in *Erichsonia attenuata*. The head is quadrate in outline, except in *Chiridotea*. The eyes are present and usually lateral, but may not be conspicuous. The antennulæ are four-jointed and similar in form throughout the family; they may or may not surpass the head in length, but are usually short and small. The basal segment of the antennulæ is more or less enlarged and usually subquadrate; the second segment is clavate; the third longer and less distinctly clavate; the fourth, or terminal, segment, corresponding with the flagellum of the antennulæ, is nearly straight along its outer, or in the natural position posterior, margin, while the opposite margin is gently curved from near the base, and rounds over more sharply at the tip; along this margin, especially toward the tip, are tufts of short setæ at regular intervals, indicating an approach toward segmentation. The antennæ have a five-jointed peduncle, varying little in form throughout the family; the first of these segments is short; the second is much larger and deeply notched on its under side; the third, fourth, and fifth segments are longer, but more slender and cylindrical or somewhat clavate. The flagellum of the antennæ may be articulated with many or few segments; it may consist of a single segment, or may be rudimentary. The maxillipeds are operculiform and cover the other parts of the mouth below. They consist, on each side, of a large semi-oval plate, with a straight interior margin, meeting its fellow of the opposite side, and bearing on this margin a short, curved, styliform organ. They are provided at the tip with stout pectinate setæ, and along the basal portion of the outer margin lies, on each side, the large external lamella. The palpi of the maxillipeds are flattened and ciliated along their inner margins, and the number of segments may be reduced to three by the coalescence of the last two and of the preceding two. The maxillæ vary but little in the family; the second or outer pair bear as usual three delicate ciliated plates; the first or inner pair are armed with stouter setæ and spines. The mandibles are robust, acutely toothed at the apex, armed with a more or less powerful molar process, and are destitute of palpi.

The thoracic segments are distinct and subequal in length, but may differ considerably in width, and are not united with the head nor with the pleon. The legs, except in the genus *Chiridotea*, are nearly similar in form throughout, and, in the first three pairs at least, are terminated by a prehensile or subprehensile hand, formed by the more or less complete flexion of the dactylus upon the propodus. The first pair of legs is usually shortest and has a triangular carpus. The anterior three pairs of legs are, in general, directed forward, and the posterior four pairs are directed backward and are less perfectly, or not at all, prehensile, a distinction that reaches its highest development in *Chiridotea*. The seventh pair of legs are absent in the young taken from the incubatory pouch, and do not generally attain quite as large size as the sixth pair.

The pleon, seen from above, consists in great part, or entirely, of a large, convex, usually pointed, scutiform piece, representing the consolidated terminal segments. As many as four of the anterior segments may, however, be more or less completely separated by articulations or indicated by lateral incisions or sutural lines. Underneath, the pleon is provided with a structure peculiar to and characteristic of this family, and the next, viz, a two-valved operculum, formed by the specially modified uropods,* or appendages of the terminal segment, closing like a pair of cupboard doors and protecting the delicate pleopods, which are lodged in a vaulted chamber excavated in the under surface of the pleon. This operculum consists, on each side, of an elongated basal plate, often strongly vaulted, angulated externally near the base, where it is articulated with the terminal segment of the pleon, and bearing at the tip one, or sometimes two, small lamellæ. One of these lamellæ usually disappears, but two are present in *Chiridotea*, as also in the foreign genera *Cleantis* and *Chætilia*. When both are present the opercular plates differ only in proportion from the ordinary form of uropods, consisting of a basal segment and two rami. Within the cavity enclosed by the opercular plates lie the usual five pairs of pleopods, each consisting of a basal segment

---

*In the last edition of the Encyclopædia Britannica (vol. vi, p. 641), these organs are described as the "anterior" abdominal appendages. They are anterior only in position, being in fact the appendages of the posterior segment.

supporting two lamellæ, and two or more of the anterior pairs are ciliated with fine plumose hairs. The inner lamella of the second pair of pleopods bears, in the adult males, a slender style articulated near the base of the inner margin and varying in length and structure in the different genera and species. The pleopods, besides their branchial office, are also of importance in locomotion, being used for swimming, which is a frequent mode of progression in this family, and is often performed with the back downward.

The females are usually broader than the males and carry their eggs and young in a pouch, on the under surface of the thorax, formed of four pairs of plates, attached to the coxal segments of the second, third, fourth, and fifth pairs of legs, and overlapping along the median line.

The known Isopoda of this family on the coast may be most easily recognized by the presence, underneath the pleon, of a two-valved operculum, opening like a pair of cupboard doors, and by the first three pairs of legs being more or less prehensile. Our genera may be distinguished by means of the following table:

```
Flagellum   ⎧articulated; legs ⎧dissimilar, last four pairs not prehensile............CHIRIDOTEA, p. 337
of the      ⎨                  ⎨                              ⎧evident above.........IDOTEA, p. 341
antennæ     ⎪                  ⎩alike, prehensile; epimera....⎩not evident above..SYNIDOTEA, p. 350
            ⎨not articulated, clavate...........................................ERICHSONIA, p. 354
            ⎩short and rudimentary.................................................EPELYS, p. 357
```

Chiridotea Harger.

Chiridotea Harger, Am. Jour. Sci., III, vol. xv, p. 374, 1878.

First three pairs of legs terminated by prehensile hands, in each of which the carpus is short and triangular, the propodus is robust and the dactylus is capable of complete flexion on the propodus; antennæ with an articulated flagellum; head dilated laterally; abdominal operculum vaulted, with two apical plates.

The two species of this genus found on our coast agree further in the following particulars: The body is short, the length being only about twice the breadth, and the outline of the head and thorax together is subcircular. The anterior part of the lateral margin of the head is produced and deeply lobed, the eyes thus appearing dorsal instead of lateral; posteriorly the head is deeply received into the first thoracic segment. The antennulæ are proportionally large, equaling or surpassing the peduncle of the antennæ. The external lamella of the maxillipeds (see pl. IV, figs. 18 and 21) is large and broad and the palpus consists of only three segments, of which, however, the last two are each composed of two coalesced segments, that are separate in the European *Ch. entomon*. Of the two segments thus formed, the terminal is quadrate or rhomboid in outline, with rounded angles and is smaller than the preceding, which expands distally toward the articulation between the two.

The thorax is deeply excavated, in front for the head and behind for the abdomen, so that the thoracic segments are much longer at the sides than along the back, when measured parallel with the axis of the animal. The epimera are separated by sutures, except in the first segment, and have their posterior angles acute. The first three pairs of legs have the dactylus capable of complete flexion upon the propodus, which is more or less swollen and supported by the short triangular carpus. In the last four pairs of legs the three corresponding segments are nearly cylindrical and the dactylus is incapable of complete flexion on the propodus.

The pleon, or abdomen, is convex throughout and pointed at the tip, and is composed, apparently, of five segments, of which the first three are separated by complete sutures, but the last two are united in the dorsal region, the sutures separating them being visible only at the sides. The opercular plates consist, on each side, of an elongated, vaulted, and attenuated plate, regularly rounded at the anterior end, truncate at the apex, and bearing just within the apex, on the inner side of the organ when closed, two ciliated, ovate or triangular plates. Of these the internal plate, or the one next the median line is much smaller than the outer; the outer also overlaps the inner, a disposition similar to that which prevails in the branchial plates or pleopods. The basal plate of the operculum is ciliated along its anterior and inner margin with bristles, which are plumose except in the region nearly opposite the articulation of the plate, where they become stouter and spine-like. The stylet on the second pair of pleopods in the males is long and slender, more than twice the length of the lamella to which it is attached.

Chiridotea cœca Harger (Say).

 Idotea cœca Say, Jour. Acad. Nat. Sci. Phil., vol. i, p. 424, 1818.
 Hitchcock, Rep. Geol. Mass., p. 564, 1833. (*I. cœca?*)
 Gould, Rep. Geol. Mass., 2d ed., p. 549, 1835; Invert. Mass., p. 337, 1841.
 Edwards, Hist. nat. des Crust., tom. iii, p. 131, 1840.
 Guérin, Iconog., Crust., p. 35, 1843.
 Dekay, Zool. New York, Crust., p. 42, 1844.
 White, List Crust. Brit. Mus., p. 94, 1847.
 Verrill, This Report, part i, p. 310 (16), 1874.
 Harger, This Report, part i, p. 569 (275); pl. v, fig. 22, 1874.
 Chiridotea cœca Harger, Am. Jour. Sci., III, vol. xv, p. 374, 1878; Proc. U. S. Nat. Mus., 1879, vol. ii, p. 159, 1879.

PLATE IV, FIGS. 16–19.

This species is at once distinguished from the following by its larger size and short antennæ, which surpass the antennulæ but little, if at all. Among the other known Isopoda of the New England coast, it may be recognized by the broad, subcircular thorax, joined with an articulated flagellum of the antennæ and a two-valved abdominal operculum. The eyes are, moreover, light-colored and inconspicuous, whence the name.

The head is but slightly excavated in front for the bases of the antennæ, and there is a more or less open notch at the sides extending nearly to the eyes. The antennulæ (pl. IV, fig. 17 *a*) are longer than the peduncle of the antennæ and have the second segment strongly clavate; the third cylindrical; the last with about a dozen tufts of short setæ; the peduncular segments are bristly, as are also those of the antennæ. The first segment of the antennæ (pl. IV, fig. 17 b) is very short, the second about three times as long, longer than any of the following segments; the third is longer and more slender than the fourth, which is nearly as broad as long; the fifth, or last peduncular, segment is more slender than any of the preceding, slightly clavate, about twice as long as broad, and longer than any except the second. The flagellum slightly exceeds the last two peduncular segments in length and consists usually of about seven segments, each bearing a tuft of short hairs near its extremity, except the first, which is much the longest, bears two such tufts, and is, apparently, composed of two segments united.

The breadth of the thorax is greater than its length along the median line. The first pair of legs (pl. IV, fig. 18 b) are a little shorter than the next two pairs, and the propodus or penultimate segment is a little more swollen. The carpus becomes slightly more elongated in the next two pairs. The last four pairs of legs are alike in form and increase in size to the sixth pair, which is the largest. The legs are bristly hairy, especially on the ischial, meral, and carpal segments, where they are provided with stout setæ curved at the tip. The basal segments bear longer and more slender plumose hairs. The epimera are ciliated on their external margins as are the lateral borders of the head and first thoracic segment and the tip of the pleon.

The operculum (pl. IV, fig. 18 c) is also ciliated with very fine hairs along its postero-external margin; the larger of the apical plates is broader than in the following species, the width being to the length as 6 to 10. The stylet on the second pair of pleopods in the male (pl. IV, fig. 19 b) considerably surpasses the cilia and is curved and acute at the tip. Adult males and females seem to be comparatively rare, and a common form of the second pair of pleopods (pl. IV, fig. 19 a) presents an acute stylet, imperfectly separated from the lamella and but slightly surpassing it in length, strongly ciliated like the lamella on its margin.

Length 12–15^{mm}; breadth 6–8^{mm}. The color in life is variable but usually dark grayish, much like the wet sand in or on which it is commonly found. It may be more particularly described as usually of a dark leaden gray on the top of the thorax, sometimes with a central spot, which may be bright pea-green, probably from the contents of the digestive cavity showing through. This dark color is continued in an arrow-shaped, or halberd-shaped, spot occupying most of the upper surface of the head. At the sides of the head and body is a mottling of light yellowish gray, darker again on the edge. The under surface of the body and the legs are pale and generally uniform in color. In alcohol the colors usually fade to a uniform straw color, with fine blackish dots, which are less conspicuous in life.

According to Say this species extends as far south as Florida. It is common on sandy beaches at many localities on the coast of New England, as at New Haven! and other localities on Long Island Sound!,

Vineyard Sound!, Nantucket!, Provincetown!, and Nahant, Mass.! It appears to be very rare, or perhaps does not occur in the northern part of the Gulf of Maine, where it is replaced by the next species; it reappears, however, on the coast of Nova Scotia, having been collected at low water by the U. S. Fish Commission in 1877, at Halifax!. It is usually found on sand below high tide, or burrowing just under the surface, but also swims with facility.

Specimens examined.

Number.	Locality.	Fathoms.	Bottom.	When collected.	Received from—	Specimens.		Dry. Alc.
						No.	Sex.	
1944	New Haven .... Vineyard Sound, Mass	....	Sand ....do	.... —, 1871	.... U. S. Fish Com.	00	♂ ♀	Alc. Alc.
1945	Off Nantucket	....	....	Sept. 8, 1875	....do	1	....	Alc.
1946	Provincetown, Mass.	....	Sand	—, 1872	....do	1	....	Alc.
1947	Nahant, Mass	l. w.	....	....	A. E. Verrill ..	3	....	Alc.
1948	Halifax, N. S	L. w.	....	—, 1877	U. S. Fish Com.	....	....	Alc.

Chiridotea Tuftsii Harger (Stimpson).

Idotea Tuftsii Stimpson, Mar. Inv. G. Manan, p. 39, 1853.
Verrill, Proc. Am. Assoc., 1873, p. 362, 1874; This Report, part i, p. 340 (46), 1874.
Harger, This Report, part i, p. 569 (275), 1874.
Chiridotea Tuftsii Harger, Am. Jour. Sci., III, vol. xv, p. 374, 1878; Proc. U. S. Nat. Mus., 1879, vol. ii, p. 159, 1879.

PLATES IV AND V, FIGS. 20–23.

This species is distinguished from the preceding by its smaller size and longer antennæ, which are about twice as long as the antennulæ and bear a slender flagellum. The eyes are also more conspicuous than in *Ch. cæca.*

The head is excavated in front above the bases of the antennæ; and the incision in the produced lateral margin is nearly closed by the overlapping of the anterior lobe. The antennulæ (pl. V, fig. 23 *a*) are slender and do not surpass the peduncle of the antennæ, the second segment as well as the third is cylindrical, and the last segment bears about nine tufts of short hairs; the peduncular segments bear also a few bristles. The antennæ (pl. V, fig. 23 *b*) have the first segment short; the second, third and fourth about equal in length and more than twice as long as the first; the fifth as long as the third and fourth together, but more slender and cylindrical; the flagellum longer than the peduncle, composed of about twelve segments and tapering from the base. The maxillipeds (pl. IV, fig. 21) have the external lamella (*c*) longer than broad.

The first pair of legs (pl. V, fig. 23 *c*) are somewhat less robust than in *Ch. cæca.* They are a little shorter than the second and third pairs, and have a much more robust hand. The fourth and succeeding pairs of legs (pl. V, fig. 23 d) are much as in the preceding species but less spiny and with a greater proportion of plumose hairs.

The external apical plate of the operculum (pl. V, fig. 23 e) is slender and twice as long as broad. The stylet on the second pair of pleopods in the males (pl. IV, fig. 22 s) does not surpass the cilia, is dilated towards the tip and obtusely pointed.

Length 9^{mm}; breadth 4.5^{mm}. The color is usually light reddish brown, speckled with darker, or marked with dark transverse patches, or bands. A specimen obtained during the summer of 1879, from a clear sandy bottom in 17 fathoms, Stellwagen's Bank, is thus described from life by Professor Verrill: "Color whitish, more or less speckled with salmon on the sides above, the specks more regular and distinct on the head, some lines and specks of flake-white on the middle of the back above the greenish stomach; base of telson salmon brown, its posterior half white; legs marked with salmon."

Dr. Stimpson's specimen "was dredged on a sandy bottom in 10 fathoms off Cheney's Head" in the Bay of Fundy. It occurs in Long Island Sound, where a specimen was taken by Dr. T. M. Prudden off New London! in 1872. The species was, however, considered rare on the coast until 1878, when it was taken in considerable abundance in Gloucester Harbor,! Massachusetts Bay, in seven to eight and a half fathoms, sand and red algæ. It has also been collected at Casco Bay,! Maine, in 1873; at low water in Prince's Cove,! Eastport, in the Bay of Fundy, in 1872, and at Halifax, N. S.,! in 18 to 25 fathoms, sand, September 5, 1877; a single specimen in each case. Three additional specimens were obtained in 1879, as detailed below.

Specimens examined.

Number.	Locality.	Fathoms.	Bottom.	When collected.	Received from—	Specimens. No.	Sex.	Dry. Alc.
1953	Off New London			—, 1872	T. M. Prudden	1	♂	Alc.
	Gloucester Harbor, Massachusetts Bay.	8½	Sand	—, 1878	U. S. Fish Com.	10		Alc.
	...do...	7½	...do...	—, 1878	...do...	00		Alc.
	Stellwagen's Bank	17	Coarse sand	Sept. 6, 1879	...do...	1	♀	Alc.
	Off Boston Harbor	16	Speckled sand	Sept. 13, 1879	...do...	2	♂♀	Alc.
803	Casco Bay, Me		Sand	July 12, 1873	...do...	1		Alc.
1952	Bay Fundy, Prince's Cove.	L. w.	...do...	—, 1872	...do...	1	♀	Alc.
1951	Halifax, outer harbor.	18–25	...do...	Sept. 5, 1877	...do...	1		Alc.

Idotea Fabricius.

Idotea Fabricius, Suppl. Ent. Syst., p. 297, 1798.

Flagellum of the antennæ articulated; legs all terminated by a prehensile hand; epimeral sutures evident above except in the first thoracic segment; pleon composed apparently of four segments, of which the last two are consolidated in the dorsal region; operculum with a single apical plate.

The species to which I propose to limit the name *Idotea** may be briefly characterized as above, and, of these, the three found on our coast agree further as follows: The body is elongated, its length being from three to four times its breadth, and the sides are nearly parallel. The head is quadrate and not produced at the sides. The eyes are lateral. The antennulæ are small and short, hardly surpassing the third segment of the antennæ. The basal segment of the antennæ is very short; the second segment much larger and deeply incised on its under surface; the third, fourth and fifth segments increase in length but decrease in diameter; the flagellum is more or less distinctly articulated, the number of articulations increasing with age. The palpus of the maxillipeds is four-jointed, the last segment being composed of two segments united, as is indicated by a notch near the tip.

The thorax is moderately arched, with the sides but little dilated in the males, somewhat more so in the females. The epimera are conspicuous and separated from their segments by a suture above, except in the first segment, but may not occupy its entire lateral margin. The legs differ but little in form throughout, being all more or less perfectly prehensile, but in the first pair only is the carpus triangular.

The pleon or abdomen appears, when seen from above, to consist of four segments, of which the first two are separated by complete sutures, but the third and fourth by sutures at the sides only. The uropods, forming the abdominal operculum, consist on each side of a flattened, elongated plate, with the anterior end rounded, the sides nearly parallel for most, or all, of its length and bearing at its truncated apex a much shorter more or less tapering or triangular plate. Neither of these plates is strongly ciliated in our species, but a stout, densely plumose bristle springs from the basal plate, on the inside, near the outer end of the articulation between the two plates. The stylet on the second pair of pleopods of the males is not elongated and may not surpass the lamella to which it is attached. The incubatory pouch is conspicuous in the females.

Our representatives of this genus may be recognized among the other known Isopoda of the coast by the following characters: The pleon appears to consist of four segments, the first three short and the third united, in the dorsal region, to the large, more or less vaulted, terminal segment; underneath the pleon is the conspicuous two-valved operculum and, in the antennæ, the flagellum consists of several segments. The three species may be distinguished by the form of the tip of the pleon, which is more or less tridentate in *I. irrorata* (p. 343), pointed in *I. phosphorea* (p. 347), and truncate in *I. robusta* (p. 349).

---

* The orthography adopted is that of Fabricius, the author of the genus.

Idotea irrorata Edwards (Say).

> *Idotea entomon* Leach, Edinb. Encyc., vol. vii, (Am. ed., p. 243, pl. ccxxi, fig. 7), "1813-14"; Trans. Linn. Soc., vol. xi, p. 364, 1815 (*not Oniscus entomon.* Linné.)
> Templeton, Lond. Mag. Nat. Hist., vol. ix, p. 92, 1836.
> Moore, Charlesworth's Mag. Nat. Hist., vol. iii, p. 294, 1839.
> *Stenosoma irrorata* Say, Jour. Acad. Nat. Sci., vol. i, pp. 423, 444, 1818.
>> Hitchcock, Rep. Geol. Mass., p. 564, 1833.
>> Gould, Rep. Geol. Mass., 2 ed., p. 549, 1835; Invert. Mass., p. 338, 1841.
>> Dekay, Zool. New York, Crust., p. 43, pl. ix, fig. 42, 1844.
> *Idotea tricuspidata* Desmarest, Dict. des Sci. nat., tom. xxviii, p. 373, pl. 46, fig. 11, 1823; Consid. Crust., p. 289, pl. 46, fig. 11, 1825.
>> "Roux, Crust. Medit., t. 29, f. 11, 12," (B. & W.)
>> Latreille, Regne Anim., t. iv, p. 139, 1829.
>> Gould, Rep. Geol. Mass., 2 ed., p. 549, 1835 (*tricuspidata* ?).
>> Edwards, Hist. nat. des Crust., tom. iii, p. 129, 1840.
>> Œrsted, Naturhist. Tidssk., B. iii, p. 561, 1841.
>> Zaddach, Crust. Pruss. Prod., p. 10, "1844."
>> Lucas, Expl. Algérie, tom. i, p. 60, 1849.
>> White, List Crust. Brit. Mus., p. 94, 1847; Brit. Crust. Brit. Mus., p. 65, 1850; Pop. Hist. Brit. Crust., p. 225, pl. 12, fig. 2, 1857.
>> Hope, Cat. Crost. Ital., p. 26, 1851.
>> Lilljeborg, Öfvers. Vet.-Acad. Förh., Årg. 9, p. 11, 1852 (*Idothea*).
>> M. Sars, Chr. Vid. Selsk. Forh., 1858, p. 151, 1859 (*Idothea*.)
>> Bate, Rep. Brit. Assoc., 1860, p. 225, 1861.
>> Norman, Nat. Hist. Trans. Northumb., vol. i, p. 25, 1865; Rep. Brit. Assoc., 1866, p. 197, 1867; op. cit., 1868, p. 289, 1869.
>> G. O. Sars, Reise ved Kyst. af Christ., 1865, p. (28), 1866 (*Idothea*).
>> Heller, Verh. zool.-bot. Ges. Wien, B. xvi, p. 728, 1866 (*Idothea*).
>> Marenzen, Arch. Naturges., Jahrgang xxxiii, B. 1, p. 360, 1867.
>> Bate and Westwood, Brit. Sess. Crust., vol. ii, p. 379, figure, 1868.
>> "Saenger, Fauna of Baltic, Imp. Soc. Nat. Sc. Mosc., viii, 1869."
> "Münter und Buchholz, Carcin. Fauna Deutschlands, 1869."
>> Czerniavski, Zoog. Pont. Comp., pp. 83, 129, "1870."
>> Metzger, J. B. Naturhist. Ges, Hannover, vol. xx, p. 32, 1871; Nordseefahrt der Pomm., 1872-'73, p. 285, 1875.
>> Möbius, Die Wirbellosen Thiere der Ostsee, p. 121, 1873. Ann. Mag. Nat. Hist., IV, vol. xii, p. 85, 1873.
>> Parfitt, Trans. Devon. Assoc., Sess. Crust., p. (19), 1873.
>> Bos, Bijd. ken. Crust. Hed. Nederl., pp. 34, 67, 1874.
>> M'Intosh, Ann. Mag. Nat. Hist., IV, vol. xiv, p. 273, 1874.
>> Stebbing, Jour. Linn. Soc., vol. xii, p. 148, 1874.
>> Catta, Ann. Sci. nat., Zool., VI, tome iii, p. 30, 1876.
>> Stalio, Cat. Crost. Adriatic, p. 206, 1877.
>> Lenz, Wirbellos. Thiere, Trave. Bucht, p. 15, 1878.
> *Idotea Basteri* Audouin, Descr. Savigny's Egypt, Crust., pl. 12, fig. 6, "1830."
>> Guérin. Iconog., Crust., p. 32, pl. xxxi, fig. 1, 1829-43.
>> "Roux, Crust. Mediterr., t. 29, f. 1-10," 1830 (B. & W.).
>> "Rathke, Fauna der Krimm, p. 380," 1830 (Edw.).
> "*Idotea variegata* Roux, Crust. Mediterr.. pl. 30, fig. 1-9," 1830 (B. & W.).
> *Idotea* (*pelagica* ?) Latreille, Cours d'Ent., Atlas, p. 12, pl. xviii, figs. 20-30, 1831.
> "*Armida bimarginata* Risso, Hist. nat. Eur. merid., 5, 109" (B. & W.).
> *Idotea irrorata* Edwards, Hist. nat. des Crust., tome iii, p. 132, 1840.
>> White, List Crust. Brit. Mus., p. 94, 1847.
>> Stimpson, Mar. Inv. G. Manan. p. 39, 1853.
>> Leidy, Jour. Acad. Nat. Sci. Phil., II, vol. iii., p. 150, 1855.

Idotea irrorata—Continued.

 Harger, This Report, part i, p. 569 (275), pl. v, fig. 23, 1874; Proc. U. S. Nat.
  Mus., 1879, vol. ii, p. 160, 1879.
 Verrill, Am. Jour. Sci., III, vol. vii, pp. 134, 135, 1874; Proc. Amer. Assoc.,
  1873, pp. 369, 371, 373, 1874; This Report, part i, p. 316 (22), 1874.
 Whiteaves, Am. Jour. Sci., III, vol. vii, p. 217, 1874; Further Deep-sea
  Dredging, Gulf of St. Lawrence, p. 15, " 1874."
 Idothea tridentata Rathke, Fauna Norw., Nov. Act. Acad., B. xx, p. 21, 1843 (*I.
  tridentata* Latreille?).
 Grube, Ausflug nach Triest, p. 126, 1861.
 ? *Idotea tricuspis* DeKay, Zool. New York, Crust., p. 42, pl. 9, fig. 35, 1844.
 Oniscus Balthicus (*Idotea marina*) Dalyell, Powers of the Creator, vol. i, p. 228,
  pl. lxiii, figs. 5–9, 1851 (*O. Balthicus* Pallas?).
 Oniscus (*Idotea*) *entomon* Dalyell, op. cit. vol. i, p. 229, pl. lxiii, fig. 10, 1851
  (*not O. entomon* Linné.).
 Idothea pelagica, M. Sars, Chr. Vid. Selsk. Forh., 1858, p. 151, 1859 (*not* of Leach).
 " *Idotea acuminata* Eichwald, Fauna Caspio-Caucasia, p. 232–233, tab. xxxvii, fig.
  6, 1842" (Czerniavski).
 Idothea balthica Meinert, Crust. Isop. Amph. Dec. Daniæ, pp. 21, 228, etc.,
  "1877" (*Oniscus Balthicus* Pallas?).

PLATE V, FIGS. 24–26.

Adults of this species are at once distinguished from the other species of the genus on our coast by the tridentate abdomen, or pleon, and young individuals, which often resemble *I. phosphorea*, may be distinguished by the epimeral sutures, which extend quite across the second and succeeding thoracic segments. For character separating them from the other Isopoda of the coast, see at the close of the generic description.

The body is smooth, not tubercular nor roughened. The head is nearly square, narrowing but slightly behind. The eyes are small. The antennulæ (pl. V, fig. 25 *a*) are short, hardly surpassing the third segment of the antennæ. The flagellum of the antennæ (pl. V, fig. 25 *b*) is longer than the peduncle, distinctly articulated, slender, and composed of from twelve to sixteen segments in the adults. When reflexed it reaches the third thoracic segment. The external lamella (*l*) of the maxillipeds (pl. V, fig. 26 *a*) is about twice as long as broad, and is obliquely truncated.

Thorax with the external margins, as seen from above, forming in the adults, a pretty regular curved line, the segments being marked by incisions instead of by serratures as in the other species. In the second and third, as well as in the posterior segments, this margin is formed wholly by the epimera.

The first three segments of the pleon terminate in acute teeth at the sides. The fourth, or last segment, has its lateral margins straight, and is more or less tridentate at the tip, the middle tooth being much the largest. In the operculum (pl. V, fig. 25 *c*) the basal plate is about three times as long as the terminal one, which is broadly truncate at the apex. The stylet (*s*) on the second pair of pleopods in the males (pl. V, fig. 26 *b*) is usually shorter than, or, in smaller specimens, about as long as the lamella to which it is attached, and is abruptly bent toward the lamella at the apex and very obliquely truncated. It is minutely serrulate toward the tip on the side opposite the lamella.

The males of this species sometimes attain a length of 30^{mm} to 38^{mm}, with a breadth of 8^{mm} to 9^{mm} but the females are smaller, rarely, if ever, exceeding 20^{mm} in length, with a breadth of 6.5^{mm}, and are found with eggs when not over 7.5^{mm} in length. The color varies greatly. Frequently it is of a nearly uniform light or dark green, or brownish with minute blackish punctations. It is often longitudinally striped with light color, or nearly white on a dark background, and the stripes may be marginal only, or accompanied, especially in the males, by a median dorsal stripe. More rarely the colors are arranged transversely in bands or blotches, and specimens thus marked are easily mistaken for the next species. The females are usually darker than the males, and often with a light lateral stripe, which may be very narrow or broken into a series of blotches.

A comparison of specimens from both sides of the Atlantic does not seem to furnish any characters by which to separate this species from the common European form, *I. tricuspidata* Desm., and as Say's trivial name has priority I have adopted it. *I. tridentata* Rathke appears to be the same species, but *I. tridentata* Latreille * is described by that author as having antennae as long as the body; further, Desmarest, just before his original description of *I. tricuspidata* says: "M. Latreille fait observer que cette idotée [*I. entomon*] est bien différente de celle que M. Leach a décrite sous le même nom, * * * * cette dernière qu'il nomme Idotée tricuspide," &c. It would not therefore appear that Latreille was at that time aware that this species had a name, much less that he had himself named it *I. tridentata*. Again, in his Cours d'Entomologie, where he copies figures, doubtless of this species, from Savigny's Egypt, he applies to them the name *Idotea* (*pelagica*?), not recognizing them as his own species. Bate and Westwood quote *I. tridentata* Latreille as a synonym of *I. tricuspidata* Desm., and their quotation † appears intended to refer to a work nearly twenty years older than that of Desmarest. They do not, however, give their reasons for deviating from the ordinary rules of priority, but, perhaps, considered as sufficient the authority of Edwards, who does the same thing. Edwards' description of *I. tricuspidata* Desm. contains, moreover, an evident error, the species being placed in a section of the genus which he thus describes: "§ 2 Espèces dont l'abdomen se compose de trois articles parfaitement distincts (le second étant composé de deux anneaux soudés ensemble sur le milieu du dos, mais séparés par une scissure sur les côtés)." *I. irrorata* is included in the same section, but under a subsection, thus correctly characterized: "*aa* Le second article de l'abdomen simple; le troisième offrant près de sa base une fissure de chaque

---

* Gen. Crust. et Ins., tome i, p. 64, 1806.

† Brit. Sess. Crust., vol. ii, p. 380. The quotation reads, "*Idotea tridentata* Latreille, Con. Crust. et Ins. 1, p. 64," and was doubtless intended for Gen. Crust. et Ins., [tome] i, p. 64, [1806].

côté." No species of *Idotea* that I have seen has the second segment of the pleon composed of two segments, united along the back but separated by an incision at the sides, as described in the parenthesis above, and two certainly of the other species included by Edwards in the section with *I. tricuspidata* agree with it in the structure of the pleon as described in *I. irrorata*. Meinert unites this species with *I. pelagica* Leach under the name *I. Balthica* (Pallas), and in this he may be right, but not being able to consult Pallas' work, I have preferred to use the earliest name that I could certainly connect with the species, rather than to introduce further confusion by adopting a name of the applicability of which I could not satisfy myself. M. Sars also regarded *I. pelagica* Leach as synonymous with *I. tricuspidata*, and says it is found as far north as Tromsoe and southward to the Mediterranean, from which statements I conclude that he intended the present species.

This species is found along the whole coast of New England! and extends southward along the coast of New Jersey at least as far as Great Egg Harbor! and northward to Nova Scotia! and the Gulf of St. Lawrence, where it has been collected by Mr. J. F. Whiteaves. From Cape Cod southward it is abundant, but toward the north it is, mostly replaced by *I. phosphorea*. It is commonly found among sea-weed along the rocky shores of bays and sounds or among the rocks, where its variety of colors affords it protection. It is also found far from land, attached to floating sea-weed, and was thus taken by Professor S. I. Smith and the writer on George's Banks!, September 14 and 15, 1872, at about 41° N. lat., 65° W. lon. One of these specimens was quite large, measuring 38^{mm} in length, but most of them were of moderate size or small. Young individuals are often taken at the surface. According to European authors it is common on the shores of Great Britain and Ireland (B. & W.); on all the shores of the North Sea (Metzger *et al.*); (*I. pelagica*) as far north as Tromsoe (M. Sars); in the Baltic, the Mediterranean, the Adriatic (Heller, Stalio, *et al.*), the Black (Czerniavski *et al.*) and the Caspian ("Eichwald") Seas, and, as with us, is of variable color and varies also somewhat in the shape of the termination of the pleon, which is, however, more or less three-toothed.

Specimens examined.

Number.	Locality.	Fathoms.	Bottom.	When collected.	Received from—	Number of Specimens.	Dry. Alc.
1078	Fire Island Beach, L. I.			—, 1870	S. I. Smith	50	Alc.
1079	do			—, 1870	do	9	Glyc.
1954	New Haven, Conn.			Nov. —, 1874	A. E. Verrill	1	Alc.
1955	Stony Creek, Conn.			Oct. 23, 1874	do	60	Alc.
1958	Lyme, Conn.				D. C. Eaton	2	Alc.
1963	Long Island Sound, off Saybrook, Conn.	4	Sand	Aug. 3, 1874	U. S. Fish Com.	2	Alc.
1964	Off Stonington, Conn.	5	Sand and gravel	Aug. 14, 1874	do	12	Alc.
1959	Noank Harbor, Conn.		Surface	July 13, 1874	do	30	Alc.

Specimens examined—Continued.

Number.	Locality.	Fathoms.	Bottom.	When collected.	Received from—	Number of specimens.	Dry. Alc.
1960	Noank Harbor, Conn..		Eel-grass	Aug. 28, 1874	U. S. Fish Com.	3	Alc.
1961	Fisher's Island			—, 1874	...do	6	Alc.
1962	Watch Hill, R. I.			Oct. —, 1872	D. C. Eaton.	2	Alc.
1965	Vineyard Sound Mass.,	Sf.		—, 1875	U. S. Fish Com.	1	Alc.
1966	...do			—, 1875	...do	7	Alc.
2153	...do			Oct. 24, 1875	...do	00	Alc.
1968	Provincetown, Mass...	L. w.		—, 1872	Smith & Harger	2	Alc.
	...do		Shore	Aug. —, 1879	U. S. Fish Com.	00	Alc.
	...do	L. w.		Aug. —, 1879	...do	00	Alc.
	...do		Eel-grass	Aug. —, 1879	...do	00	Alc.
	...do	Sf.		Sept. 4, 1879	...do	10	Alc.
426	Beverly, Mass				A. E. Verrill...	8	Alc.
	Gloucester, Mass		Tide-pool	—, 1878	U. S. Fish Com.	2	Alc.
	Gloucester, Mass.,	7-10	Sand, red algæ.	—, 1878	...do	00	Alc.
	Outer Harbor.						
	Between Bass Island and Matinicus Rocks.			—, 1873	Capt. G. H. Martin.	5	Alc.
	Casco Bay, Me			—, 1873	U. S. Fish Com	11	Alc.
1975	Casco Bay, Ram I...	L. w.		—, 1873	...do	4	Alc.
2150	George's Bank	Sf.		Sept. —, 1872	Smith & Harger	6	Alc.
1977	Bay of Fundy	L. w. & sf.		—, 1872	U. S. Fish Com.	2	Alc.
1978	Off Halifax, N. S.			—, 1877	...do	1	Alc.
1979	Nova Scotia	L. w.		—, 1877	...do	1	Alc.
	Durham coast, England				Rev. A. M. Norman.	4	Alc.
	St. Vaast, la Hogue				Jardin des Plantes.	1	Alc.

Idotea phosphorea Harger.

Idotea phosphorea Harger, This Report, part i, p. 569 (275), 1874; Proc. U. S. Nat. Mus., 1879, vol. ii, p. 160, 1879.

Verrill, Am. Jour. Sci., III, vol. vii, pp. 43, 45, 131, 1874; Proc. Amer. Assoc., 1873, pp. 362, 367, 369, 1874; This Report, part i, p. 316 (22), 1874.

Whiteaves, Am. Jour. Sci., III, vol. vii, p. 218, 1874; Further Deep-sea Dredging, Gulf of St. Lawrence, p. 15, "1874."

PLATE V, FIGS. 27–29.

This species may be distinguished from the others on this coast by the pointed abdomen or pleon. Young individuals sometimes resemble the young of *I. irrorata*, but may still be distinguished by the epimeral sutures of the second and third thoracic segments, which do not entirely cross the segment, but allow more or less of the posterior part of the edge of the segment to form a part of the margin of the animal as seen from above. From *Synidotea nodulosa* it may be distinguished by the evident epimeral sutures and by the three acute teeth at the base of the pleon on each side, instead of a single obtuse tooth, as in that species. For characters separating it from the other Isopoda of the coast see at the close of the description of the genus.

The body, especially of the young, is rough and tubercular along the median line and often also laterally. Older specimens are much smoother, losing their large median tubercles but never becoming as smooth as in the preceding species. The head is narrowed behind. The eyes are of moderate size. The flagellum of the antennæ (pl. V, fig. 28 a) is shorter than the peduncle, and consists of about ten to fourteen segments. The maxillipeds (pl. V, fig. 28 b) have the external lamella (l) broader than in the preceding species, with its inner margin straight and its outer margin curving pretty regularly to a slightly attenuated tip.

The epimera of the second, third, and fourth pairs are rounded behind, and those of the last three pairs are less acute than in *I. irrorata*.

Pleon ovate, a little constricted near the middle and pointed, its three proximal segments rather less acute than in the preceding species. The basal plate of the operculum (pl. V, fig. 28 c) tapers toward the end, and the terminal plate is triangular, a little longer than broad. The stylet on the second pair of pleopods in the male (pl. V, fig. 29 s and s') is slender, nearly straight, surpasses the lamella to which it is attached, and is obliquely truncate.

Length 25^{mm}; breadth 7^{mm}. The color is very varied, usually dark green or brownish, with patches of yellow or whitish, transversely or obliquely arranged. I have never observed a striped pattern of coloration, so common in *I. irrorata*, and it must occur very rarely if at all. The color is usually darker than in that species.

This species is found associated with the last among rocks and seaweed along the entire coast of New England! and extends northward to Halifax!, Nova Scotia, and the Gulf of St. Lawrence!. It appears to be a more northern species than *I. irrorata*, as it is comparatively rare south of Cape Cod, while it is abundant in Casco Bay, Maine, and in the Bay of Fundy.

Specimens examined.

Number.	Locality.	Fathoms.	Bottom.	When collected.	Received from—	Specimens.		Dry. Alc.
						No.	Sex.	
1080	South End, New Haven, Conn.			Nov. —, 1874	A. E. Verrill			Alc.
1081	Stony Creek, Conn.			Sept. 23, 1874	do			Alc.
1083	Off Saybrook, Conn.	4	Sand	Aug. 3, 1874	U. S. Fish Com.			Alc.
1084	Long Island Sound			Aug. 10, 1874	do	3	y.	Alc.
1085	South Fisher's Island.	9		Aug. 21, 1874	do	1		Alc.
1087	Vineyard Sound, Mass.			—, 1871	do			Alc.
1088	do			—, 1872	do			Alc.
2147	do				do	1		Alc.
1990	Off Nantucket	15		Sept. 8, 1875	do	1	y.	Alc.
	Cape Cod Bay	14	Green mud.	Sept. 15, 1879	U. S. Fish Com.	2		Alc.
	do	7	Coarse yellow sand.	Sept. 15, 1879	do	3		Alc.
144	Nahant, Mass.				A. E. Verrill	3		Alc.
	Gloucester, Mass.	7–10	Sand and algæ	—, 1878	U. S. Fish Com.	00	♂♀ y.	Alc.
	Ten Pound Island, Gloucester, Mass.			—, 1878	do			Alc.
	Annisquam Mass.	½		—, 1878	A. Hyatt	20		Alc.
1091	Casco Bay, Me.	8f.		Aug. 4, 1873	U. S. Fish Com.	1	y.	Alc.
1092	do			—, 1873	do	4		Alc.
1093	Casco Bay, Ram Island.	L. w.		—, 1873	do	4		Alc.
1094	Bay of Fundy			—, 1872	do	1	♀	Alc.
	Bay of Fundy, Whiting River.			—, 1872	do	00	♂♀	Alc.
	Off Halifax, N. S.	18		—, 1877	do	2		Alc.
	Egmont Bank, Gulf of St. Lawrence.			—, 1873	J. F. Whiteaves.	1		Alc.

Idotea robusta Kröyer.

? *Idotea metallica* Bosc, Hist. nat. des Crust., tom. ii, p. 179, pl. 15, fig. 6, 1802.
Idothea robusta Kröyer, Naturhist. Tidsskr., II, B. ii, p. 108, 1846; Voy. en Scand., Crust., pl. 26, fig. 3, "1849."
Reinhardt, Grönlands Krebsdyr, p. 35, 1857.
Stimpson, Proc. Acad. Nat. Sci., 1862, p. 133, 1862.
Verrill, Am. Jour. Sci., III, vol. ii, p. 360, 1871; This Report, part i, p. 439 (145), 1874 (*Idotea*).
Harger, This Report, part i, p. 569 (275), pl. v, fig. 24, 1874; Proc. U. S. Nat. Mus., 1879, vol. ii, p. 160, 1879 (*Idotea*).
Lütken, Crustacea of Greenland, p. 150, note, 1875.

PLATE VI, FIGS. 30-32.

This species is easily recognized within the genus by the pleon, which is broadly truncate at the apex and not at all pointed. The pleon is also large and more swollen above than in the other species. For characters separating it from other Isopoda, see near the close of the generic description.

The entire upper surface, except perhaps that of the pleon, is somewhat rugose. The head is nearly square, with the eyes large and prominent. The antennæ (pl. VI, fig. 31 *a*) have the second segment large, the flagellum short, usually of less than ten articulations. Under a sufficient power these organs are seen to be clothed with a very fine close pubescence, which also occurs in a less degree upon the legs. The maxillipeds (pl. VI, fig. 32 *a*) have the external lamella (*l*) short and oval.

The legs are robust and spiny. The epimera, projecting, give a serrated appearance to the sides of the thorax, as seen in figure 30, plate VI, and the dorsum is more convex than in the other species.

The pleon is large and convex, its sides are nearly parallel beyond the middle, and it is broadly truncate, or even somewhat emarginate, at the apex. The basal plate of the operculum (pl. VI, fig. 31 *c*) is elongated, with parallel sides; the terminal plate less than one-fourth as long and nearly square, but tapering slightly and somewhat broader than long. The male stylet on the second pair of pleopods (pl. VI, fig. 32 *c*, *s*) reaches the end of the lamella, to which it is attached, and is slightly curved and rounded at the tip.

Length of male 28^{mm}; female 22^{mm}; breadth 9^{mm}. Color bright blue or green above when alive, becoming darker and dull in alcohol, without the markings of the other species, but often with metallic reflections, when seen in the water, where it is commonly taken swimming free or among masses of floating sea-weed.

It is thus found in mid-ocean, and was described by Kröyer from specimens taken in about 60° north latitude between Iceland and Greenland. It was taken in considerable abundance at Fire Island Beach!, on the south shore of Long Island, by Professor S. I. Smith in 1870; also by the U. S. Fish Commission at Vineyard Sound!, Mass., often in company with *I. irrorata* Edw.; at George's Banks!, September, 1872, small specimens, 5^{mm} in length; between Boon Island and Matinicus Rocks, near the

Isles of Shoals!, by Capt. G. H. Martin, of the schooner 'Northern Eagle,' in 1878, and at Halifax!, Nova Scotia, by the U. S. Fish Commission in 1877, whence it extends to at least 60° north latitude.

The figure and description of *Idotea metallica* given by Bosc correspond well with small specimens of this species such as were taken by Professor S. I. Smith and the writer on George's Banks, and the locality he gives, "the high seas," corresponds also with the habit of this species, so that I am inclined to think that his name ought to be restored. I have, however, retained Kröyer's name, since he so thoroughly described and so well figured the species as to leave no doubt of its identity.

Specimens examined.

Number.	Locality.	Habitat.	When collected.	Received from—	Specimens. No.	Sex.	Dry. Alc.
1080	Fire Island Beach, Long Island	Surface	—, 1879	S. I. Smith	46	♂♀	Alc.
1398	Vineyard Sound, Mass	Surface	—, 1875	U. S. Fish Com	1		Alc.
1999	...do	Surface	...do	...do	00	♂♀	Alc.
2002	...do	Surface	July 14, 1871	...do	2		Alc.
2003	...do	Surface	Oct. 24, 1875	V. N. Edwards	00		Alc.
2004	...do	Surface	Nov. 16, 1875	...do	00		Alc.
2000	George's Bank	Surface	Sept. —, 1872	Smith & Harger	4	y.	Alc.
2001	Halifax, N. S.	Surface	—, 1877	U. S. Fish Com	1		Alc.

Synidotea Harger.

Synidotea Harger, Am. Jour. Sci., III, vol. xv, p. 374, 1878.

Antennæ with an articulated flagellum; epimeral sutures not evident above; pleon apparently composed of two segments, united above but separated at the sides by short incisions; operculum with a single apical plate; palpus of maxillipeds three-jointed.

Of the two species that I had referred to this genus I had been able to examine only the first when this paper was placed in the hands of the printer. Two specimens of the second species were collected during the summer of 1879, and an examination of their characters leaves no doubt of their generic affinity. Except in the particulars above specified the description already given of the genus *Idotea* will in general apply also to the present, but the species are characterized by a firmer and more solid structure, the segments being more closely articulated and the integument having a somewhat shelly appearance. The pleon is further consolidated than in that genus, the only trace of its composite nature, as seen from above, being a slight incision on each side near the base and running up somewhat obliquely toward the dorsal surface. The well-developed and distinctly articulated flagellum of the antennæ serves easily to distinguish the species from those of the following genera of the family.

Synidotea nodulosa Harger (Krøyer).

Idothea nodulosa Krøyer, Naturhist. Tidssk., II, B. ii, p. 100, 1846; Voy. en Scand., Crust., pl. 26, fig. 2, 1849.
Reinhardt, Grønlands Krebsdyr, p. 34, 1857.
Lütken, Crust. Greenland, p. 150, "1875."
Synidotea nodulosa Harger, Am. Jour. Sci., III, vol. xv, p. 374, 1878; Proc. U. S. Nat. Mus., 1879, vol. ii, p. 160, 1879.

PLATE VI, FIGS. 34–35.

This species may be recognized most easily by the pleon, which is entire, except for a slight incision near the base on each side, and tapers to a blunt but not at all bifid point. The articulated flagellum of the antennæ distinguishes it from *Erichsonia*.

The head and body are roughened and tubercular, having a prominent median row of tubercles and coarse rugæ along the sides of the thorax. The head has a median notch in front, and immediately above this a prominent tubercle directed forward, and succeeded on the median line by two less prominent tubercles. In front of each eye is a still larger tubercle, directed forward and projecting over the anterior margin of the head; behind and within, there are two smaller oval tubercles. The eyes are large, convex, and very prominent. The peduncular segments of the antennæ (pl. VI, fig. 34 b) increase gradually in length from the first and decrease in diameter from the second, which lacks the lateral incision seen in *Idotea*. The flagellum is distinctly articulated, with about nine segments, of which the last two are very minute. The maxillipeds (pl. VI, fig. 35 a) have the external lamella (*l*) of an irregular shape, emarginate on the inner side and obtusely pointed. The outer maxillæ (pl. VI, fig. 35 b) are armed on their external lobe with strong, curved, pectinated setæ, which become much elongated and stout at the tip of the lobe. The inner maxillæ (pl. VI, fig. 35 c) resemble these organs in other members of the family.

The first four thoracic segments have their external margins rounded. In the last three the margins are more nearly straight, but with rounded angles. The first pair of legs (pl. VI, fig. 34 c) are much shorter than the second, and the propodus in the first pair is bristly on what is, in the ordinary position, the upper side.

The pleon is short, and tapers from the base. It is convex, bears two or three small tubercles on the median line near the base, and an impressed transverse line in continuation of the short lateral incisions. The basal plate of the operculum (pl. VI, fig. 34 d) is oblique at the base with rounded angles, and is somewhat vaulted, with an oblique elevation extending from the articulation to the inner distal angle. The inner margin is straight, and the outer parallel with it to near the end. The terminal plate is slightly oblique at the base, and is elongated triangular, about twice as long as broad. The free margins are finely ciliated, except at and near the base, and the inner margin of the basal plate bears also scattered stouter hairs. The stylet of the males on the second pair of pleopods (pl. VI, fig. 35 d, s) is longer and stouter than in any of our species of *Idotea*. It is nearly twice the length of the lamella, to which it is attached, and of an elongated spatulate form tapering to an obtuse point. The lamellæ are provided with but few cilia, which extend less than half the way from the end of the lamella to the end of the stylet.

Length 10.5^{mm}.; breadth 3.5^{mm}. Females proportionally broader; length 8.^{mm}; breadth 3^{mm}. Color in alcohol gray, often with brownish transverse markings.

This species seems to agree with *Idotea nodulosa* Kröyer, from Southern Greenland, as described and figured, except that the epimeral sutures are not evident above; the lateral margins of the segments are, however, somewhat thickened and prominent with rugæ, as shown in his figure, and I have no doubt that it is the same as his species. It was dredged off Halifax! by the Fish Commission at several localities in the summer of 1877, in from 16 to 190 fathoms on sandy and rocky bottoms, with red algæ at one locality. A specimen was brought from George's Banks! by Mr. Joseph P. Schemelia, of the schooner 'Wm. H. Raymond,' in the summer of 1879, and Mr. J. F. Whiteaves has sent to the Museum for examination two specimens collected by Mr. G. M. Dawson, in 111 fathoms, Dixon Entrance!, north of Queen Charlotte Island, British Columbia. The range of the species would therefore be, as at present known, from George's Banks to Greenland and the Arctic Seas, and southward on the Pacific coast as far as British Columbia..

Specimens examined.

Number.	Locality.	Fathoms.	Bottom.	When collected.	Received from—	Specimens. No.	Sex.	Dry. Alc.
	Dixon Entrance, Q. C. I.	111			J. F. Whiteaves	2		Alc.
2006	Off Halifax, N. S.	16	Stones, sand, red algæ.	—, 1877	U. S. Fish Com.	2		Alc.
2007	South of Halifax, 120 miles.	100	Gravel and pebbles.	Sept. 1, 1877	....do	1	♀	Alc.
2008	Halifax, outer harbor	18	Sand, stones	Sept. 4, 1877	....do	2		Alc.
	....do	16	Rocks, nullipore	Sept. 4, 1877	....do	1		Alc.
	George's Banks			—, 1879	J. P. Schemelia	1	♀	Alc.

Synidotea bicuspida Harger (Owen).

Idotea bicuspida Owen, Crustacea of the Blossom, p. 92, pl. xxvii, fig. 6, 1839.
Streets and Kingsley, Proc. Essex Inst., vol. ix, p. 108, 1877.
Idotea marmorata Packard, Mem. Bost. Soc. Nat. Hist., vol. i, p. 296, pl. viii, fig. 6, 1867.
Whiteaves, Further Deep-sea Dredging in Gulf of St. Lawrence, p. 15, 1874.
Idotea pulchra Lockington, Proc. Cal. Acad. Sci., vol vii, p. 45, 1877.
Synidotea bicuspida Harger, Proc. U. S. Nat. Mus., 1879, vol. ii, p. 160, 1879.

This species may be most easily recognized among the known Isopoda of our coast by the form of the pleon, which is nearly triangular in shape, marked by a slight incision at each side near the base, and distinctly bicuspid at the tip.

The body is rather more robust than in the last species, the length being only about two and a half times the breadth, and is peculiarly marked above by depressed and mostly curved lines, varying in length but mostly short, and confined principally to a region on each side of the median line and extending across the head but not the pleon.

The head is broadly emarginate in front, with a median notch, and its antero-lateral angles are prominent. The eyes are at the widest part of the head, and are strongly convex. The posterior outline of the head is nearly in the form of three sides of a hexagon. The antennulæ attain about the middle of the fourth antennal segment. The antennæ are about one-half as long as the body. The first two antennal segments are short and apparently articulated so as to admit of but little motion; the third segment is a little longer than the first two taken together, and is the largest of the antennal segments in diameter; the fourth segment is somewhat longer than the third, and the fifth or last peduncular segment is the longest, and is followed by a flagellum, a little shorter than the peduncle and composed of about fourteen segments. The last three peduncular segments of the antennæ are somewhat bristly hairy. The maxillipeds are nearly as in the preceding species. The outer maxillæ are destitute of the elongated, pectinate setæ found in that species.

The thoracic segments vary but little in length measured along the median line, but the fifth, sixth, and seventh are slightly shorter than the preceding ones, and this difference is still greater measured along the margins of the segments, where the first is longest, the next three about equal, and the last three shorter. The legs are robust, the first pair shortest, and all more or less bristly hairy. The lateral margins of the segments are much less rounded than in *S. nodulosa*.

The pleon is short, the length being scarcely greater than the breadth at base; above, it is nearly smooth, the impressed lines, so conspicuous in the lateral region of the thorax, being continued for but a slight distance upon its surface. The incision at each side near the base is continued upward and forward by a depressed line on each side; the lateral margins are gently convex to near the tip, which is distinctly bicuspid. The basal plate of the operculum is traversed obliquely by a longitudinal ridge on the external surface, and is rounded in front, slightly narrowed behind, and bears a short, triangular, terminal plate, its length being but little greater than its breadth.

Length 15.5^{mm}; breadth 6^{mm}. Color in alcohol grayish, with white cloudings. Lockington says: "When recent, the coloration of this species is very beautiful, consisting of red cloudings on a lighter ground."

There seems to be no doubt in regard to the synonymy of this species as published by Streets and Kingsley, adopted by the writer in a previous publication, and given above.

The only specimens that I have examined were two, brought from the Grand Banks!, in the summer of 1879, by Mr. Charles Ruckley, of the schooner 'Frederick Gerring, jr.', Capt. Edwin Morris. Dr. Packard's locality is "Sloop Harbor, Kynetarbuck Bay [Labrador], seven fathoms on a sandy bottom." Whiteaves records the species from Orphan Bank, Gulf of Saint Lawrence. Lockington's specimens were collected on the "west coast of Alaska, N. of Behring's Strait, by W. J. Fisher, naturalist of the U. S. S. Tuscarora, Deep-Sea Sounding Expedition." Owen's locality is "the Arctic Seas."

Erichsonia Dana.

Erichsonia Dana, Am. Jour. Sci., II, vol. viii, p. 427, 1849.

Antennæ six-jointed, the terminal or flagellar segment not articulated, clavate; palpus of the maxillipeds four-jointed; legs all nearly alike, prehensile or sub-prehensile; pleon with its segments consolidated into a single piece.

This genus is represented within our limits by two well-marked species, which further agree in the following characters: The head is quadrate, with the eyes lateral. The antennulæ are short, not surpassing the third segment of the antennæ. The antennæ are well developed, more than half as long as the body, with a very short basal segment articulated with little or no motion to the second segment, which is two or three times as long as, and of greater diameter than the first. It is, as usual in the family, incised at its distal end on the under surface. The next three segments are nearly cylindrical. The last or flagellar segment is the longest, and is slightly clavate.

The legs are all terminated by a prehensile or sub-prehensile hand, the dactylus being capable of considerable or complete flexion on the more or less swollen propodus. This flexion is most complete in the first pair. The first two pairs of legs arise near the anterior margin of the segments to which they belong. The place of attachment to the segment moves gradually backward in the following pairs until the last two pairs arise near the posterior margin of the last two segments. The epimera are more or less evident from above, at least in the last two segments.

The pleon constitutes about one-third the length of the body, and is consolidated into a single piece; it bears a more or less evident tooth on each side near the base, and is dilated and obtusely triangular at the apex. The basal plate of the operculum is oblique at the anterior end and abruptly narrowed posteriorly, where it bears a densely plumose bristle, as in *Idotea*; the terminal plate is triangular. The stylet on the second pair of pleopods in the males is well developed, surpassing the cilia; it is minutely denticulated or spinulose near the end and very acute.

The two species found on our coast have but a slight external resemblance to each other, and may be distinguished at a glance, as will be seen from the specific descriptions, and from the figures (pl. VI, fig. 36, and pl. VII, fig. 38). The long, clavate terminal segment of the antennæ distinguishes them at once from young specimens of *Idotea*, especially *I. phosphorea*, which sometimes resemble *E. filiformis*. This character of the antennæ serves, indeed, to distinguish the two unlike representatives of the present genus from all the other Isopoda of our coast.

Erichsonia filiformis Harger (Say).

> *Stenosoma filiformis* Say, Jour. Acad. Nat. Sci., vol. i, p. 424, 1818.
> Edwards, Hist. nat. des Crust., tom. iii, p. 134, 1840.
> Dekay, Zool. New York, Crust., p. 44, 1844.
> *Idotea filiformis* White, List Crust. Brit. Mus., p. 95, 1847.
> *Erichsonia filiformis* Harger, This Report, part i, p. 570 (276), pl. vi, fig. 26, 1874; Proc. U. S. Nat. Mus., 1879, vol. ii, p. 160, 1879.
> Verrill, This Report, part i, p. 316 (22), 1874.

PLATE VII, FIGS. 38–41.

This species may be at once distinguished from the following by the strongly serrated outline of the sides, as seen from above. The clavate terminal segment of the antennæ distinguishes it from the other known Isopoda of our coast.

The body is slender and elongated, but less so than in the next species, the sides are nearly parallel and there is a median row of prominent tubercles, one, large and bifid, on the head, and one upon each thoracic segment. The eyes are prominent. The antennulæ (pl. VII, fig. 39 *a*) surpass the middle of the third antennal segment. The first segment of the antennæ (pl. VII, fig. 39 *b*) is very short; the terminal segment is bristly hairy toward the apex. The external lamella of the maxillipeds (pl. VII, fig. 41 *a*) is emarginate on the outer side toward the apex.

The thoracic segments each bear a prominent median tubercle near their posterior margins, and the first bears also a smaller tubercle near its anterior margin. In the first two segments the posterior external angles are salient and much elevated. The angulated epimera are evident from above in front of these projections. In the third and fourth segments both lateral angles are salient but not elevated. In the last three segments, only the anterior angles are produced, but the epimera fill the places of the posterior angles. This arrangement gives the appearance of fourteen teeth upon each side of the thorax, and the prominent divergent tooth on the pleon makes, in all, fifteen.

The operculum (pl. VII, fig. 39 *d*) is a little more vaulted than in the next species and shorter; the basal plate is less than three times as long as broad; the terminal plate is triangular. The stylet on the second pair of pleopods in the male (pl. VII, fig. 41 *b, s*) is slightly curved, finely spinulose near the apex on the side toward the lamella, and minutely and sharply denticulate on the opposite side at the apex, as shown in the enlarged figure (*s'*) of the distal portion of the stylet.

Length 11^{mm}; breadth 3.4^{mm}. The color is a usually dull neutral tint without bright markings, but sometimes more or less variegated with brown or reddish, fading in alcohol.

This species was originally described from Great Egg Harbor, New Jersey, where Say found it in company with *Idotea irrorata*. It is not uncommon along the shores of Long Island Sound! and as far east as Vineyard Sound, Mass.! but has not yet been found north of Cape Cod. It is usually found in tide-pools or among eel-grass and algæ, and has been taken from a depth of 7 fathoms.

Specimens examined.

Number.	Locality.	Fathoms.	Bottom.	When collected.	Received from—	Number of specimens.	Dry. Alc.
2010	Long Island Sound					00	Alc.
2011	Thimble Islands				A. E. Verrill		Alc.
2012	Long Island Sound, Fisher's Island Sound.	7	Sand and shells	—, 1874	U. S. Fish Com.	1	Alc.
2013	Long Island Sound	4½	Sand and gravel	—, 1874	...do	2	Alc.
2014	...do			—, 1874	...do	1	Alc.
2015	...do			Sept. 10, 1874	...do	2	Alc.
2016	Noank		Eel-grass	—, 1874	...do	2	Alc.
2017	Vineyard Sound			—, 1875	...do	2	Alc.

Erichsonia attenuata Harger.

Erichsonia attenuata Harger, This Report, part i, p. 570 (276), pl. vi, fig. 27, 1874; Proc. U. S. Nat. Mus., 1879, vol. ii, p. 160, 1879.
Verrill, This Report, part i, p. 370 (76), 1874.

PLATES VI and VII, FIGS. 36 and 37.

This species is at once distinguished from the preceding by its slender form and regular outline; the clavate antennal flagellum distinguishes it from other Isopoda.

The body is smooth throughout and about six times as long as broad, without prominent irregularities and narrowly linear in outline. The eyes are small and black. The antennulæ (pl. VII, fig. 37 *a*) are short, slightly surpassing the second antennal segment. The antennæ (pl. VII, fig. 37 *b*) are stout and smoother than in the preceding species. The external lamella of the maxillipeds (pl. VII, fig. 37 *c*, *l*) is oval and regularly rounded at the tip.

The thoracic segments increase in size to the third, which is equal to the fourth, and the last three are of a gradually decreasing size. The epimera are nowhere conspicuous, but may usually be seen from above, especially in the posterior segments.

The pleon presents only slight traces of a lateral tooth near its base and is but little dilated toward the tip. The operculum (pl. VII, fig. 37 *d*) is longer than in the preceding species, the basal plate is more than three times as long as broad, the terminal plate elongated triangular and obtuse. The male stylet on the second pair of pleopods (pl. VII, fig. 37 *e*, *s*) is nearly straight, hardly surpasses the cilia, and is minutely denticulated near the acute apex.

Length 15mm; breadth 2.5mm. Alcoholic specimens are of a light grayish yellow, with minute black punctations.

It was abundant in eel-grass at Great Egg Harbor, New Jersey! in April, 1871, and has also been found at Noank, Conn.! on eel-grass, but is not common. It has not been found north of Cape Cod.

Specimens examined.

Number.	Locality.	Fathoms.	Bottom.	When collected.	Received from—	Specimens.		Dry. Alc.
						No.	Sex.	
1226	Great Egg Harbor, N. J.	......	Eel-grass	Apr. —, 1871	S. I. Smith	60		Alc.
2018	Noank, Conn	......	do	—, 1874	U. S. Fish Com.	1	♀	Alc.

Epelys Dana.

Epelys Dana, Am. Jour. Sci., II, vol. viii, p. 426, 1849.

Antennæ shorter than the antennulæ and with only a rudimentary flagellum; palpus of the maxillipeds three-jointed; legs all terminated with prehensile hands; pleon consolidated into a single segment with a basal lobe on each side.

Two small and closely allied species from this coast have been referred to this genus. They resemble each other very closely and may be at once recognized by their depressed ovate form, very short antennæ, and generally dirty appearance. The form of the body and absence of powerful mandibles distinguish them from the male *Gnathia*. The length of the body is between two and three times its width. It is marked by a depressed line on each side, running from the posterior part of the head, across the thoracic segments, nearer to their lateral margins than the median line, except perhaps in the last segment, thence continued to inclose a prominent hemispherical protuberance on the anterior part of the pleon, giving the animal somewhat the appearance of a trilobite. The body is slightly roughened under a lens, or sometimes minutely hirsute. The head is slightly dilated at the sides, with the anterior angles produced, and bears a pair of broad, low, triangular tubercles on its anterior part, and a curved posterior depression. The eyes are lateral and prominent, the antennulæ are longer than the head, surpass the antennæ, and have the basal segment but little enlarged. The antennæ (pl. VIII, fig. 45 *b*) are shorter than the head, not surpassing the third antennular segment, the segments increasing in length to the fourth; fifth as long as the fourth, but more slender, bearing a minute, slender rudiment of a flagellum, which is setose at the tip.

The thoracic segments have thick evident margins; first segment smallest, somewhat embracing the head; third and fourth largest;

last segment curving around the base of the pleon. The epimera are not evident from above. The legs (pl. VIII, fig. 46 a) are slender and all terminated by a slender prehensile hand, of which the finger, or dactylus, becomes almost acicular in some of the posterior pairs. All the legs are more or less hairy.

The pleon bears on each side, near its base, a rounded lobe, which is separated from the large posterior portion by a more or less evident incision. Dorsally it is convex, and presents two hemispherical elevations, the proximal more convex than, but only half as large as, the distal. They are separated by a broad and deep groove, and the distal convexity is continued upon the obtusely-pointed apex of the pleon. The operculum (pl. VIII, fig. 46 b) is vaulted; its basal plate is rounded anteriorly, carinate near its inner margin, contracted externally for the distal third of its length and truncate at the tip, where it bears a stout elongated-triangular finely ciliated terminal piece. The basal plate is coarsely ciliated on its inner margin, and bears a few plumose hairs along its outer free margin. The stylet on the second pair of pleopods in the males is short and stout, surpasses the lamella but not the cilia, and is spinulose just below the blunt apex.

Both species are of a dull neutral color, and commonly covered with particles of mud or other foreign matter. They occur on piles, or under stones, in muddy places, and are dredged on muddy bottoms.

Epelys trilobus Smith (Say).

    Idotea triloba Say, Jour. Acad. Nat. Sci. Phil., vol. i, p. 425, **1818**.
    Edwards, Hist. nat. des Crust., tome iii, p. 134, 1840.
    Dekay, Zool. New York, Crust., p. 43, 1844.
    Leidy, Jour. Acad. Nat. Sci., II, vol. iii, p. 150, 1855.
    Jaera? *triloba* White, List Crust. Brit. Mus., p. 97, 1847.
    Epelys trilobus Smith, This Report, part i, p. 571 (277), pl. vi, fig. 28, 1874.
    Verrill, Am. Jour. Sci., III, vol. vii, p. 135, 1874; Proc. Amer. Assoc., 1873, p. 372, 1874; This Report, part i, p. 370 (76), 1874.
    Harger, Proc. U. S. Nat. Mus., 1879, vol. ii, p. 160, 1879.

PLATE VII, FIGS. 42 and 43.

This species may be recognized among our Isopoda by its appearance when seen from above, recalling the form of the trilobites, the flattened dorsal surface being marked, as in those animals, by two lateral longitudinal depressions. The pleon is consolidated into a single piece and the antennæ have only a rudimentary flagellum. It closely resembles the next species, but is smaller and most readily distinguished by the lateral margin of the thorax, which is, especially in the anterior part, nearly even instead of zigzag from the projecting angular segments. The anterior angles of the head are also less produced.

The pleon is shorter and broader, its breadth being to its length as six to ten. The deep transverse groove across the pleon is continued to the margin, with only, at the most, traces of a tubercle at each side. The stylet on the second pair of pleopods of the male (pl. VII, fig. 42 b, *s*, and *s'*), is a little less elongated than in the next species, not attaining the middle of the cilia.

Length 6^{mm}; breadth 2.3^{mm}. The color is uniform, dull, usually obscured by the adhering particles of dirt.

This species was described by Say from Egg Harbor!, New Jersey, where specimens were also collected by Professors Verrill and Smith, in April, 1871, among eel-grass. It has also been found at Savin Rock!, near New Haven, and Noank Harbor!, on piles and among eel-grass; at Vineyard Sound!; Mass., at Provincetown!, Mass., near Cape Cod in 1879; sparingly near Gloucester! Mass., in 1878, and even as far north as Quahog Bay!, about thirty miles northeast of Portland, Me., where it was taken by the United States Fish Commission, in 1873, along with *Venus mercenaria* and other southern forms.

Specimens examined.

Number.	Locality.	Fathoms.	Bottom.	When collected.	Received from	Number of specimens.	Dry. Alc.
1227	Great Egg Harbor, N. J.		Eel-grass	Apr. —, 1871	S. I. Smith	7	Alc.
2019	Savin Rock, New Haven.	L. w.		1871-1872	A. E. Verrill	2	Alc.
2020	Noank Harbor			Aug. 12, 1871	U. S. Fish Com	12	Alc.
2021	...do			July 13, 1871	...do	7	Alc.
2022	...do		On piles	July 27, 1874	...do	1	Alc.
2024	...do		Eel-grass	—, 1874	...do	60	Alc.
2023	Watch Hill, R. I			Apr. —, 1873	A. E. Verrill	4	Alc.
2025	Vineyard Sound			—, 1871	U. S. Fish Com.	2	Alc.
2026	...do			—, 1871	...do	1	Alc.
2027	...do	L. w.	Sand	—, 1871	...do	2	Alc.
	Provincetown, Mass.	L. w.		Aug. —, 1879	U. S. Fish Com	2	Alc.
	...do	½	Eel-grass	—, 1879	...do	1	Alc.
	...do		Shore	—, 1879	...do	9	Alc.
	Gloucester, Mass		Tide-pools	—, 1878	...do	2	Alc.
2028	Quahog Bay, Me	L. w.	Muddy	—, 1873	...do	3	Alc.

Epelys montosus Harger (Stimpson).

Idotea montosa Stimpson, Mar. Inv. G. Manan, p. 40, 1853.
Epelys montosus Harger, This Report, part i, p. 571 (277), 1874; Proc. U. S. Nat. Mus., 1879, vol. ii, p. 161, 1879.
Verrill, Am. Jour. Sci., III, vol. vii, p. 45, 1874; Proc. Amer. Assoc., 1873, p. 367, 1874; This Report, part i, p. 370 (76), 1874.
Smith and Harger, Trans. Conn. Acad., vol. iii, p. 3, 1874.
Whiteaves, Further Deep-sea Dredging, Gulf St. Lawrence, p. 15, "1874."

PLATE VIII, FIGS. 44-47.

This species closely resembles the preceding, and may be recognized among our Isopoda by the characters mentioned under the former species, from which it is distinguished by the following characters: The eyes are prominent; the anterior angles of the head salient. The tubercles on the head are more prominent than in the former species. The lateral margins of the thoracic segments, especially the second, third, and fourth, are angulated and salient. The pleon is more elongated than in the last species, its breadth being to its length as 5.5 to 10, and the depression crossing it is partially interrupted at each side by a tubercle which often projects, as seen from above, just behind the basal lobe, forming a shoulder to the large terminal lobe. The stylet on the second pair of pleopods in the males (pl. VIII, fig. 47, *s* and *s'*) attains about the middle of the cilia.

Length 10^{mm}; breadth 4^{mm}; color, as in the preceding, dull, and usually much obscured by adhering dirt.

A few specimens were collected in Whiting River, near Eastport, Maine, in 1872, which are much more decidedly hirsute than is usual, both on the upper surface and on the legs as well. In other respects they appear to be referable to this species, although the posterior thoracic segments are rather less angulated at the lateral margin. They may be worthy of a variety name *hirsutus*.

Dr. Stimpson's specimens were "taken in deep water on sandy and muddy bottoms" in the Bay of Fundy, and this species usually replaces the last in the northern localities. It has, however, been taken as far south as Block Island Sound!, near the eastern end of Long Island Sound, in 18 fathoms, sandy bottom, and in 29 fathoms Vineyard Sound!. North of Cape Cod it is more common. It was dredged in 25 fathoms on St. George's Bank!, at Stellwagen's Bank! in 20 to 40 fathoms, rocky and sandy bottom; Casco Bay!, 16 to 17 fathoms mud; Bay of Fundy!, at many localities, usually on muddy bottoms, and in 16–18 fathoms mud and stones, off Halifax!, Nova Scotia, by the Fish Commission, and in 14 fathoms off Richibucto, in the Gulf of Saint Lawrence, by Mr. J. F. Whiteaves. The greatest depth positively recorded is 29 fathoms, but it may very likely have come also from a depth of 40 fathoms near Stellwagen's Bank.

Specimens examined.

Number.	Locality.	Fathoms.	Bottom.	When collected.	Received from—	Number of specimens.	Dry. Alc.
2029	Vineyard Sound	29		Sept. 14, 1871	U. S. Fish Com	8	Alc.
2030	Block Island Sound	18	Sand	—, 1874	do	1	Alc.
	Off Boston Harbor	16	Speckled sand, shells	Sept. 13, 1879	do	2	Alc.
	Gloucester Harbor, Mass	7–8½	do	—, 1878	do	30	Alc.
2032	George's Bank	25		—, 1872		2	Alc.
2031	Stellwagen's Bank	20–40	Rocks and sand	—, 1873	A. S. Packard	1	Alc.
2033	Casco Bay	16	Mud	July 12, 1873	U. S. Fish Com	00	Alc.
2035	do	17	do	Aug. 30, 1873	do	00	Alc.
	do			—, 1873	do	00	Alc.
2038	Bay of Fundy, Eastport.			—, 1872	do	6	Alc.
2039	do			—, 1872	do	2	Alc.
2040	Bay of Fundy, Whiting R.	2	Mud	—, 1872	do	6	Alc.
2041	Seal Cove, Grand Menan.	8–10		—, 1872	do	10	Alc.
2042	Off Halifax, N. S	16	Stones, sand, red algae.	—, 1877	do	4	Alc.
2043	do	18	Mud, fine sand	Sept. 15, 1877	do	2	Alc.

VI.—ARCTURIDÆ

Form elongated; antennæ large and strong; first four pairs of legs directed forward, ciliated, last three pairs ambulatory; segments of pleon more or less consolidated; uropods operculiform.

This well marked family is as yet represented on our coast by a single species of the genus *Astacilla* Fleming, *Leachia* or *Leacia* of Johnston and other authors. The family can be easily recognized by the four anterior pairs of legs, which are directed forward and strongly ciliated on their inner margins with long slender hairs. The form of the body is elongate and may be very much so, as in our species the length of the body in the male is twenty times as great as its diameter at the middle; in the female eight times. The head is of moderate size and the eyes prominent. The four-jointed antennulæ have the basal segment large and swollen. The antennæ are large and powerful organs, approaching or even surpassing the body in length, with the first two segments short, the second deeply incised below as in *Idotea*, the next three segments elongated, and the flagellum varying in the genera, being multiarticulate in *Arcturus*, and composed of not more than four segments in *Astacilla*. The mouth parts resemble, in general, those of the *Idoteidæ*. The fourth thoracic segment is more or less elongated. The last three pairs of legs are ambulatory, differing much from the first four pairs. The segments of the pleon are more or less united, and the uropods are modified, as in the preceding family, to form an operculum for the more delicate anterior pleopods. They are wholly inferior, and consist on each side of a large basal segment, straight on the median line, where it meets its fellow of the opposite side, and bearing, in our genus, two small terminal plates at the apex.

This structure of the pleon and its appendages, together with the structure of the antennulæ, antennæ, and the parts of the mouth, point to a close relationship between this family and the *Idoteidæ*. With the *Anthuridæ*, however, with which they have often been associated, they seem to have little in common, except, perhaps, the elongate form of body. Even this feature is approached also in the *Idoteidæ*, in *Erichsonia*, for example.

Astacilla Fleming.

Leacia (*Leachia*) Johnston, Ed. Phil. Jour., vol. xiii, p. 219, 1825 (*non* Lesueur).
Astacilla Fleming, Encyc. Brit., 7th ed., vol. vii, p. 502.
Johnston, Lond. Mag. Nat. Hist., vol. viii, p. 494, 1835.

Antennal flagellum short, not more than four-jointed; fourth thoracic segment elongated, and, in the females, bearing the incubatory pouch on its inferior surface.

The characters given above seem sufficient to warrant the separation of this genus from *Arcturus*, notwithstanding the fact that the young of some species, and probably of all, have the fourth thoracic segment no longer than the others as noticed by Johnston[*], and later by Stebbing[†], who draws from the fact an argument against the validity of the genus. I fail to see, however, why the argument would not be equally valid against the use, among mammals, of characters drawn from the horns and teeth. Nothing is more common, in case of a genus or family possessing a special development of some organ or set of organs, than to find that the young of such a group resemble the adults of less specialized groups. If, however, as may be possible, a gradation can be established between forms which, like *Arcturus Baffini*, have the fourth thoracic segment large but only slightly elongated, and forms like *Astacilla longicornis* or *A. granulata*, in which this segment is much elongated, equaling or surpassing the other six in length, there would then be, perhaps, no sufficient reason for retaining both genera. For the present it seems desirable to keep them separate, and to the characters given above we may add the following:

The head is produced at the sides around the bases of the antennulæ, and is united dorsally with the first thoracic segment, the sutures being evident only at the sides where the segment is produced around the hinder part of the head. The flagellum of the antennæ consists of three, or sometimes only two, distinct segments and a terminal spine, which is perhaps to be regarded as a third or fourth segment. The maxillipeds (pl. IX, fig. 52 *a*) are robust and operculiform, with a thick external lamella and a five-jointed palpus, but little flattened. The mandibles are destitute of palpi.

The first three thoracic segments are subequal and short; the fourth much elongated in both sexes; in the males it is slender and cylindrical; in the females it is more robust, and bears on its inferior surface the incubatory pouch. This pouch is thus confined to a single segment, and is composed of a pair of elongated lamellæ, attached along their outer margins, and overlapping widely along the ventral surface. It occupies nearly the entire inferior surface of the segment. The last three thoracic segments are short and subequal, and the articulation at the posterior end of the fourth segment is capable of considerable motion, and, in our species, is usually flexed backward nearly at a right angle. The first pair of legs (pl. VIII, fig. 49 *b*) have the basis directed backward and the remaining segments ciliated and turned forward, and is more robust than the three succeeding pairs, which are slender, of nearly equal size, and consist of only five segments, which are turned forward from the basis and held beneath the head. They are strongly ciliated, especially on the last three segments. One of the fourth pair of legs is shown on plate VIII, figure 50. The last three pairs of legs are of entirely different structure, being robust and prehensile with strong short dactyli.

The pleon is consolidated into a single segment, which, however, shows traces of its composite nature. It is vaulted above and excavated on

---

[*] Lond. Mag. Nat. Hist., vol. ix, p. 81, fig. 15, 1836.
[†] Ann. Mag. Nat. Hist., IV, vol. xv, p. 187, 1875.,

its inferior surface for the delicate pleopods, which are protected by the operculiform uropods. Both rami of the uropods are present in our species, but the outer is much the larger and conceals the delicate inner ramus in an exterior view. The outer ramus only is thickened and of functional importance as an operculum.

The habits of these animals are described by Goodsir in the Edinburgh New Philosophical Journal, vol. xxxi, p. 311. He says, "With the dredge I have procured specimens * * * * alive, and have kept them in glass jars of sea-water with sand and corallines, and have thus been enabled to watch their habits closely.

"Under the circumstances just stated, each individual will select a branch of coralline, will keep that branch exclusively to itself, and will defend it with the greatest vigor against all intruders. It fixes itself to its resting-place by means of its true thoracic feet, and seldom uses these for progression. When it falls to the bottom of the vessel, it fixes its long pointed antennae firmly into the sand, and, with the assistance of the true feet, drags and pushes itself forward. This, however, may not be a natural mode of progression, but may be adopted in consequence of the artificial circumstances in which the animal is placed.

"Swimming is the natural mode of progression. It is amusing to see one of these animals resting, in an erect posture, on a branch of coralline, by means of its true thoracic feet, waving its body backwards and forwards, throwing about its long inferior antennae, and ever and anon drawing them through its anterior fringed feet, for the purpose of cleaning them. It frequently darts from its branch, with the rapidity of lightning, to seize with its long antennae some minute crustaceous animal, and returns to its resting-place to devour its prey at pleasure.

"In this manner the antennae are the only organs employed in seizing and enclosing the prey, which they drag to the anterior thoracic feet, which hold it while it is being devoured."

I have discarded Johnston's name *Leachia*, or according to his orthography *Leacia*, proposed in 1825, as being preoccupied by Lesueur * in the Mollusca in 1821. *Astacilla* is used by Fleming in the 7th edition of the Encyclopædia Britannica; 1842 is given as the date in the copy of the seventh volume of the Encyclopædia that I have seen, but Johnston refers to Fleming in 1835 as authority for the name, quoting the Encyclopædia. Fleming says in the Encyclopædia (vol. vii, p. 502): "The genus was instituted by the Rev. Charles Cordiner of Banff in 1784 for the reception of a British species which has been denominated *Astacilla longicornis*." I have not been able to find whether Cordiner published the name at that early date or whether it was a manuscript name only. If actually published in 1784 it would have many years' priority over *Arcturus*, and the author who would unite the genera should use the name *Astacilla*. Even if not published until 1835 it appears to have the best claim to recognition as the generic name of the type here treated of.

---

* Jour. Acad. Nat. Sci. Phila., vol. ii, p. 89, 1821.

Astacilla granulata Harger (G. O. Sars).

Leachia granulata G. O. Sars, Arch. Math. Nat., B, ii, p. 354 [251], 1877.
Astacilla Americana Harger, Am. Jour. Sci., III, vol. xv, p. 374, 1878.
Astacilla granulata Harger, Proc. U. S. Nat. Mus., 1879, vol. ii, p. 164, 1879

PLATES VIII AND IX, FIGS. 48–52.

The elongated fourth thoracic segment distinguishes this species at once from all the other Isopoda of our coast.

The body is in the female eight times and in the male about twenty times as long as broad, the breadth being measured across the fourth thoracic segment. It is roughened and tuberculated throughout. The head is produced at the sides in front beyond the middle of the basal segment of the antennulæ, and is tuberculated above and crossed by two transverse grooves, the first between, and the second behind the eyes, while a third similar groove evidently marks the place of the suture between the head and the first thoracic segment. The eyes are lateral, prominent, round-ovate, broadest in front. The antennulæ in the female slightly surpass the second segment of the antennæ, in the male they nearly attain the middle of the third segment, the flagellar segment being elongated in the male, longer than the three peduncular segments together (pl. VIII, fig. 48 a). The second and third segments of the antennulæ are in both sexes short and slender. The antennæ are fully three-fourths as long as the body; the first segment is shorter than that of the antennulæ, being surpassed at the sides by the lateral processes of the head and thus concealed in a lateral view; the second segment is large, scarcely longer than broad, and presents below a deep angular sinus in the distal margin, as in *Idotea*; third segment about as long as the head; fourth segment longest, slightly exceeding the fifth, which is equal to the first three taken together. The flagellum* (pl. VIII, fig. 49 a) is less than half the length of the last peduncular segment and usually consists of three distinct segments, of which the first is as long as the other two; the second is equal in length to the third, which is tipped with a terminal spine or claw, probably to be regarded as a fourth segment. Sometimes, however, only two distinct segments exist in the flagellum besides the claw. The flagellar segments are finely and sharply denticulate along the margin which is inferior when the antennæ are straightened. The character of this denticulation is shown in figure 49 a' on plate VIII, where a small section of the margin is shown enlarged 100 diameters. The maxillipeds (pl. IX, fig. 52 a) are robust and cover the other parts of the mouth; the external lamella (*l*) is ovate and in the figure is somewhat bent outward from its natural position. The palpus of the maxillipeds is five-jointed and but little flattened, strongly ciliated along the inner margin. The terminal lobe

---

* The figure of the animal (pl. VIII, fig. 48,) was sent to the engraver before I had seen any specimens except the imperfect ones collected in 1877, and the flagellum of the antennæ was dotted from the young specimens. Fig. 49 a on plate VIII was made from a specimen obtained in 1878.

(pl. IX, fig. 52 a, m') is quadrate, scarcely ciliated at the apex, and distinctly articulated with the maxilliped. The outer maxillæ (pl. IX, fig. 52 b) are three-lobed and strongly ciliated. The inner maxillæ (pl. IX, fig. 52 c) are two-lobed, the lobes robust and short, the outer armed with short spines at the apex, the inner with three slender curved setæ.

The thoracic segments are coarsely granulated or tuberculated; the first is produced at the sides around the head nearly to the eyes; the others have their anterior and posterior margins transverse. The fourth segment in the female is a little less than three times as long as broad, and is longer than the other six segments taken together, but is only four-fifths as long as the last three segments together with the pleon. It is tuberculated, especially above, but bears no prominent tubercles or spines, and is subcylindrical. In the male this segment (pl. VIII, fig. 48 b) is more elongate and much more slender, exceeding in length the three following segments with the pleon. In the ordinary position the thorax is geniculate at the posterior articulation of the fourth segment, forming nearly a right angle with the rest of the body. The last three segments have their epimeral regions angulated and salient. The first pair of legs (pl. VIII, fig. 49 b) are of moderate length and, beyond the basal segment, flattened; the basal segment is directed backward but the leg is bent upon itself at the ischium and the remaining segments are directed forward and applied to the under surface of the head. The ischium and merus support but few cilia, and these mostly along their inner margins, but the carpus, propodus, and dactylus are not only ciliated on the inner margin with slender simple cilia, but also bear on the side toward the body stout scattered spinulose setæ, which are specially abundant on the propodus. The opposite side of the leg is nearly smooth. The second, third, and fourth pairs of legs are five-jointed and similar to each other, except that the basal segments of the second and third are somewhat shorter than in the fourth (pl. VIII, fig. 50). The second pair is shorter than the third, and the fourth is a little the longest. All these legs are directed strongly forward and habitually held nearly in the position shown in the figure, under the anterior surface of the body and the head. The last three segments are furnished with elongated setæ along their inner margins. These setæ are inserted in two rows and so placed as to diverge at an open angle. The dactyli appear to be obsolete in these legs. The fifth, sixth, and seventh pairs of legs are of quite a different and more ordinary structure. They contain the full number of segments, and are terminated by robust, slightly curved dactyli. A young specimen obtained has only two pairs of legs of the ordinary form, the last or seventh pair being represented only by rounded tubercles, one on each side of the seventh segment.

The pleon is elongate-ovate, narrower in the male (pl. VIII, fig. 48 c). Dorsally it is strongly convex, especially in front. It is two-thirds as long as the fourth thoracic segment in the female, and three-fifths as long as that segment in the male. It is provided with rather coarse tubercles in front, which are arranged transversely in three rows, and behind the third row is a deep transverse groove, behind which the tubercles are less prominent and more of the character of granulations. On each side before the middle is a prominent, sub-acute tooth, directed outward and backward immediately above the articulation of the uropods. The tip of the pleon is not spiniform, but only slightly attenuated and obtuse. The pleopods are delicate in structure, and the anterior pairs are ciliated. The uropods or opercula are more than nine-tenths as long as the under surface of the pleon (pl. VIII, fig. 48 c), but cannot be seen from above. They consist on each side (pl. VIII, fig. 51) of an elongated, semi-oval, basal, lamellar segment, thickened and vaulted externally, with the anterior end rounded, and bearing a salient semi-circular process on the outer margin near the anterior end, for articulation with the pleon. Posteriorly this plate is tapering and it is broadly truncated at the tip, where it bears two lamelliform rami. Of these the external is thick, like the basal segment, and is of an elongate triangular form and completes the operculum behind, while the inner ramus is a small and delicate oval plate, articulated to the basal segment near its inner distal angle, and completely covered and concealed by the outer ramus when the operculum is closed. The inner ramus is sparingly ciliated at the tip. The pleopods are very delicate, and the anterior pairs are ciliated.

In the females the lamellæ forming the incubatory pouch are thickened and tuberculated or granulated along the outer edge where they are attached to the segment. The thickened area is bounded by a longitudinal ridge, beyond which the lamella is thin, smooth, and translucent, permitting the eggs to be seen through it, and the thin portion of the right lamella (in the specimen examined) overlaps its fellow of the opposite side so far as to bring its edge along the base of the ridge bounding the thickened portion of the opposite lamella. Near the anterior end and on the outer side is a rounded lobe in the margin of the lamella for articulation with the segment.

Length of female 10^{mm}; male 11^{mm}; diameter of fourth thoracic segment, female 1.2^{mm}; male 0.52^{mm}; color in alcohol, nearly white.

This species was described by the writer without having seen Sars' description of *Leachia granulata*. The volume containing his description has since been obtained by the Yale College Library, and a careful comparison of our specimens with his description leaves little doubt that the species is identical with his. His specimens were somewhat larger than ours, females measuring 14^{mm} and males 17^{mm}. The females in *A. longicornis* Sowerby are much larger than the males, and the reverse relation of size in this species appears to be unusual in the genus.

Specimens were first collected on this coast on George's Bank !, in the summer of 1877, and the three then obtained were found adhering to *Primnoa*, and had been dried and somewhat broken. Better specimens were collected adhering to the cable of the schooner 'Marion,' at Banquereau!, by Capt. J. W. Collins, August 25, 1878, and a fine specimen was obtained in seven fathoms off Miquelon Island!, south of Newfoundland, by Capt. C. D. Murphy and crew of the schooner 'Alice M. Williams,' July 3, 1879. Sars' specimens were collected between Norway and Iceland at stations 18 and 48, of which the respective localities as given by him are latitude 62° 44.5′ north, longitude 1° 48′ east, in 412 fathoms, clayey bottom, and latitude 64° 36′ north, longitude 10° 21.5′ west, in 299 fathoms, clay and sand.

Specimens examined.

Number.	Locality.	Fathoms.	Bottom.	When collected.	Received from—	Specimens. No.	Sex.	Dry. Alc.
2045	George's Bank			—, 1877	U. S. Fish Com	2	♀ y	Alc.
2046	....do			—, 1877	....do	1	♂	Alc.
	Banquereau, N. S.	250	Rocky	—, 1878	Capt. J. W. Collins.	3	♀	Alc.
	Off Miquelon Island.	7		July 3, 1879	Capt. C. D. Murphy and crew.	1	♀	Alc.

VII.—SPHÆROMIDÆ.

Body short and convex; head transverse; antennulæ and antennæ multiarticulate, with evident distinction into peduncle and flagellum; mandibles palpigerous; epimera united with the thoracic segments; anterior segments of the pleon short, united and articulated with the large terminal segment; uropods lateral with only one movable ramus.

This family is sparingly represented on the eastern coast of the United States, and within our limits only a single species is found, belonging to the typical genus *Sphæroma*. The animals are usually of small size, and have the body short, broad, and convex. The head is transverse, and both pairs of antennæ are inserted near together below its anterior margin. These organs are much better developed than in the following family. The epimera are faintly indicated in the thoracic segments by impressed lines. The anterior segment of the pleon is similarly marked with transverse sutures indicating the segments of which it is composed. The last segment is large, and one or more of the posterior segments may be notched, tuberculated, spiny, or variously modified, as occurs in many foreign genera. Below, the pleon is much excavated for the pleopods, which, as usual, are in five pairs, the anterior three ciliated. In the males a slender stylet is articulated near the base of the inner lamella of the second pair, and lies along its inner side, so that in the natural position they lie close together on opposite sides of the middle line of the body. These pleopods, though received into a cavity in the under surface of the pleon, are not protected by any operculum nor opercular plates, as in most of the preceding families, nor is the external pair thickened, as in the *Anthuridæ*.

Sphæroma Latreille.

Sphæroma Latreille, Hist. nat. des Crust. et des Ins., tome vii, p. 14, 1804.

Body contractile into a sphere; antennulæ and antennæ short or of moderate length; maxillipeds with a five-jointed palpus; legs all ambulatory; dactyli short and thick; uropods short, ramus and basal segment subequal.

The name of this genus is derived from the peculiar habit of many of the species of rolling themselves into a ball when alarmed. The body is so constructed as to facilitate this operation, the antennulæ and antennæ being received into a groove at the side of the head; the epimeral regions of the thoracic segments behind the first are narrowed nearly to a point and project well downward so as to meet very close together and still leave room for the included legs, while the uropods, shutting together like a pair of scissors, fold also partly under the large terminal segment of the pleon and fill the crevice between the pleon and the head. The maxillipeds in this genus are provided with a long densely ciliated five-jointed palpus. The maxillæ are much as in the *Idoteidæ*, the outer pair three-lobed and strongly ciliated, the inner two-lobed with the inner lobe small and tipped with pectinate setæ, the outer larger and armed with curved denticulated spines. The mandibles have a strong molar process, a dentigerous lamella armed with acute teeth, and a three-jointed palpus.

The legs are rather weak and nearly alike throughout, all ambulatory. The pleon is scarcely narrower than the segments of the thorax and appears to consist of two * segments only, of which the first is much like the last thoracic segment, but more strongly produced at the sides than is that segment and marked with impressed lines. It is articulated with considerable motion to the large scutiform terminal segment, which, in this genus, is rounded and entire at the tip, and not strongly tuberculated nor spiny. Anteriorly, the angles of this segment are produced downward into a rounded lobe in front of the shoulder from which arise the uropods. These organs are not greatly elongated; the basal segment is produced into a plate about equal in size to the single ramus.

Sphæroma quadridentatum Say.

Sphæroma quadridentata Say, Jour. Acad. Nat. Sci. Phil., vol. i, p. 400, 1818.
  Dekay, Zool. New York, Crust., p. 44, 1844.
  White, List Crust. Brit. Mus., p. 102, 1847.
  Harger, Am. Jour. Sci., III, v., p. 314, 1873; This Report, part i, p. 569 (275), pl. v., fig. 24, 1874; Proc. U. S. Nat. Museum, 1879, vol. ii, p. 164, 1879.
  Verrill, This Report, part i, p. 315 (21), 1874.

PLATE IX, FIG. 53.

The outline of the body when extended is a pretty regular ellipse, but the animal, when disturbed, rolls itself into a ball with facility, and by

---

* The pleon is inadvertently described by Bate and Westwood in the British Sessile-Eyed Crustacea, vol. ii, p. 404, as "having all the segments fused together."

this habit may be distinguished from the other marine Isopods of our coast.

The head is rounded in front with an elevated margin, and a slight median projection between the bases of the antennulae. The eyes are small and sub-triangular, widely separated. The antennulae and the antennae are inserted on the inferior surface of the head, and, when the animal contracts, they are received into a groove along the margin of the head and anterior thoracic segment. The antennulae (pl. IX, fig. 54 a) have the basal segment large, the second segment small and conical, the third slender, cylindrical; the flagellum about ten-jointed, ciliated, shorter than the peduncle. In the antennae (pl. IX, fig. 54 b) the peduncular segments decrease but little in diameter, and increase in length from the first to the fifth, and are followed by a flagellum about as long as the peduncle, tapering from the base, with the basal segments strongly ciliated along their inner or anterior distal margins. The antennae are separated at the base by a triangular, somewhat projecting epistome, which also partly separates the bases of the antennulae. The maxillipeds have the basal segment short and somewhat triangular, with plumose setae at the acute apex, and a five-jointed palpus, of which the first segment is short and smooth, and the following segments strongly ciliated along more or less of their inner margins. The outer maxillae are terminated by three ovate rather acute lobes, which are strongly ciliated. The inner maxillae have the inner lobe tipped with four pectinated curved setae, and the outer armed with strong denticulated spines. The mandibles are robust and bear on their external surface at the apex a dentigerous lamella, or usually two such on the right mandible, receiving the lamella of the left between them; below the lamella is a strongly ciliated ridge supporting the dentigerous lamella and connecting it with the molar process, which is large and strong. The mandibular palpi are slender, with the last segment sub-semicircular, bearing at its apex a few serrated spines, and below a comb of straight setae; the middle segment bears a similar comb with stouter spiny setae at the ends.

The first thoracic segment is longer than the others, and much elongated at the sides, embracing the head as far as its anterior margin. Above this lateral expansion on each side the segment is excavated for a projecting lobe of the head behind the eye. The second, third, and fourth segments are somewhat shorter than the first and longer than the fifth, sixth, and seventh. The margin of the last segment bends slightly backward at the middle. In the thoracic segments behind the first the epimeral sutures are indicated by a faint depressed line, below which the lateral margin of the second segment tapers to an obtusely rounded point, the third is more acutely pointed, the fourth oblique and acute behind, the fifth and sixth also oblique but less acute, and the seventh rounded. The legs are weak, hairy, and much alike throughout, formed for walking, and none of them chelate. The dactylus in all is short and robust, armed with a stout curved spine or claw at the tip, and a smaller

24 F

straight spine below it. In the first pair of legs the carpus is short and triangular, the ischium and merus bear on their upper margin a row of long slender plumose hairs. In the second and third pairs of legs these hairs are also found, and the carpus is longer. The fourth pair of legs are robust, the following pairs more slender to the seventh. All are well provided with slender hairs, with a few stouter ones intermixed.

The anterior segments of the pleon are consolidated into a single piece somewhat resembling the last thoracic segment, but marked at the sides by depressed lines, indicating sutures, as shown in pl. IX, fig. 53. At the sides this segment is broadly rounded and projects much below the seventh thoracic when the animal is contracted. The large terminal segment has a similar lobe in front of the bases of the uropods. At the insertion of the uropods the segment is considerably contracted laterally, but is rounded and strongly margined behind. Its anterior lobe, all the thoracic segments, and the head are also margined by an elevation running completely around the animal except where it is interrupted by the uropods. The uropods extend nearly to the tip of the telson, and consist on each side of a basal segment continued backward into a narrow oval plate with entire margins, flattened below, where a similarly-shaped ramus is articulated near its base, the two shutting together like the blades of a pair of scissors. The articulated plate bears four more or less acute serrations on its exterior margin, whence the specific name. The pleopods are ciliated, and the second pair (pl. IX, fig. 54e) bears, in the male, on the inner lamella, a slender curved stylet, longer than the lamella, and articulated near its base.

Length about 8""", breadth 4""". The color, as usual in shore species, is variable; some are of a uniform slaty gray, many are marked on the dorsal surface with a whitish, cream color, or rosaceous patch, bordered more or less with dark or black. This patch has commonly a longitudinal direction, and is usually symmetrical, and may be broad or much narrowed in the middle. On the dark or barnacle-covered rocks, where these animals are often found, the colors are evidently protective, but they are imperfectly preserved in alcohol.

This species was described by Say, who "found these animals very numerous on the beach of Saint Catherine's Island, Georgia, concealing themselves under the raised bark, and in the deserted holes of the *Teredo*, &c., of such dead trees as are periodically immersed." He also gives East Florida as a locality, and there are specimens in the Yale Museum from Florida! It extends as far north as Provincetown, Mass.! near the extremity of Cape Cod. It is common on the southern shore of New England!, and is usually found among algæ or rocks.

Specimens examined.

Number.	Locality.	Fathoms.	Bottom.	When collected.	Received from	Number of specimens.	Dry. Alc.
2054	Florida				Smithsonian Inst	00	Alc.
1224	Great Egg Harbor, N. J.			April, 1871	Smith & Verrill	...	Alc.
2053	New Haven, Conn				S. I. Smith	5	Alc.
	Savin Rock, New Haven	l. w.	Rocky			00	Alc.
2052	Stony Creek, Conn	l. w.	Rocky			00	Alc.
2049	Vineyard Sound, Mass			——, 1871	U. S. Fish Com.	3	Alc.
2050	...do	l. w.		——, 1871	...do	1	Alc.
2051	...do			——, 1875	...do	5	Alc.
	Provincetown, Mass	l. w.		Aug. —, 1879	...do	00	Alc.
	...do	½	Eelgrass	Aug. —, 1879	...do	1	Alc.

VIII.—LIMNORIIDÆ.

Body compressed; antennulæ and antennæ short, subequal; mandibles palpigerous, formed for gnawing; feet not prehensile, all similar, with short, robust dactyli; epimera united with the thoracic segments; pleon of six distinct segments; pleopods similar in form throughout; uropods lateral, biramous.

This family as constituted above contains the single genus *Limnoria* Leach, which appears also to contain but few, or perhaps a single, species* of wide distribution. This genus was placed in the tribe *Asellotes homopodes* with the *Asellidæ* by Edwards, without, however, having examined the animals himself. He has been generally followed in this arrangement by later authors. Previous authors had associated the genus, as it appears to me more justly, with *Sphæroma* and the *Cymothoidæ* in the wide signification of the latter term. White, in his List of British Crustacea, used the name *Limnoriadæ* to include this genus with the *Asellidæ*. I have preferred to constitute a new family for the genus, which has, however, evident relations with the *Sphæromidæ*, and perhaps should yet be united with that family.

Under the circumstances family characters can scarcely be separated with certainty from those of generic or even of specific value only, but for the purpose of comparison with other families certain important characters may be here stated. The body is somewhat depressed dorsally, but is also compressed at the sides, and when extended is subvermiform. It is nearly capable of being rolled into a ball, as in the genus *Sphæroma*. The head is of moderate size and strongly rounded above, as in *Sphæroma*, and the eyes are widely separated and on the sides of the head, a condition not usual in the *Asellidæ*. The antennulæ are short and stout and the basal segment is but little larger than the second; the flagellum

---

* It is perhaps hardly necessary to remark that *L. xylophaga* Hesse, Ann. Sci. nat., tome x, p. 101, pl. ix, 1868, is not an Isopod. According to Prof. Smith it is *Chelura terebrans* Phillipi, a boring amphipod often found associated with *Limnoria*, article by that author in the Proceedings of the U. S. National Museum, pp. 232-235.

consists of a single, almost rudimentary segment. The antennae differ widely from any in the *Asellidæ*, since they are less robust than the antennulæ, and but little longer; the peduncular segments are all short, having almost the same proportion to each other as in *Sphæroma* (see pl. IX, figs. 54 *b* and 56 *b*), the last two being together about equal in length to the first three, instead of far surpassing them as in the *Asellidæ*; the flagellum is short and few-jointed, mostly made up of a tapering basal segment, and not at all resembling the slender multiarticulate flagellum of the *Asellidæ*. The mandibles are adaptively modified in accordance with the boring habits of the species, but the other mouth parts do not seem to present characters from which comparisons need be drawn with other families.

The legs are somewhat similar to those seen in many *Asellidæ*, being furnished with short dactyli, each armed with a strong curved claw, and a shorter spine below. A similar form of leg is, however, seen in *Sphæroma*. The epimera are united to the lateral margins of the thoracic segments almost precisely as in *Sphæroma*, an arrangement that does not prevail in the *Asellidæ*.

The pleon has all its six segments well developed and perfectly separated from each other, while in the *Asellidæ* they are united into a single scutiform segment, or at most, the basal segment only is more or less distinct. The pleopods are of the normal number and similar in form and texture throughout; the anterior pairs are ciliated. Each pair of pleopods consists of a basal segment, bearing an inner narrow lamella and an outer oval one, which, except in the fifth pair, are well ciliated. In the male the inner lamella of the second pair bears, on its inner margin, a stylet, as in *Sphæroma* and many other genera of Isopoda. In the *Asellidæ* the branchial pleopods are in fewer than five pairs, and are protected in front by a simple or compound operculum of firmer texture than the other pleopods. Dr. Coldstream* fell into an error in describing the respiratory organs as consisting of "six pairs of scale-like bodies, pendant from the anterior segments of the tail, * * arranged in three rows, in an imbricated manner, one of each kind ('oval' and 'nearly quadrangular') being articulated together on a common peduncle on either side." He further describes, loc. cit., p. 324, "two vesicular bodies of an oval form" behind the branchiæ. These organs were without doubt the external lamellæ of the fifth pair of pleopods, as shown by his figure. There are, however, four instead of three ciliated pairs anterior to the last pair, one of which was overlooked by Dr. Coldstream, and in this error he has been followed by Bate and Westwood.† If the observations of Dr. Coldstream had been correct, an affinity might have been indicated with the *Asellidæ*. The terminal segment is flattened and scutiform, in shape resembling that of *Jæra*, but the uropods are strictly lateral, being attached at the broadest part of the segment and in front of the middle.

---

* Edinburgh New Phil. Journal, vol. xvi, p. 323.
† Brit. Sessile-Eyed Crustacea, vol. ii, p. 350.

The relations of the present family with the *Sphæromidæ* appear to be more close, but the structure of the mandibles and perhaps also that of the maxillipeds, the fully segmented pleon and the biramous uropods seem to be characters of family value, which, however, a fuller investigation of the boring *Sphæromidæ* might go far to break down.

Limnoria Leach.

Limnoria Leach, Edinburgh Encyc., vol. vii, p. "433" (Am. ed., p. 273), "1813-14."

Mandibles with a nearly even chisel-like cutting-edge at the tip and no molar process; maxillipeds elongate, with a well developed external lamella and a five-jointed palpus; first thoracic segment large; uropods with the outer ramus very short and almost obsolete.

The above characters differ from those by which Leach separated this genus from *Cymothoa* and the *Sphæromidæ*, with which he associated it.

Limnoria lignorum White (Rathke).

"*Cymothoa lignorum* Rathke, Skrivt. af Naturh. Selsk., v. 101, t. 3, f. 14, 1799" (White).

Limnoria terebrans Leach, Ed. Encyc., vol. vii, p. "433" (Am. ed., p. 273), "1813-14"; Trans. Linn. Soc., vol. xi, p. 374, 1815; Dict. Sci. nat., tome xii, p. 353, 1818.

Samouelle, Ent. Comp., p. 109, 1819.

Desmarest, Consid. Crust., p. 312, 1825.

Latreille, Règne Anim., tome iv, p. 135, 1829.

Coldstream, Edinb. New Phil. Jour., vol. xvi, pp. 316-334, pl. vi, 1834.

"Hope, Trans. Ent. Soc. Lond., vol. i, p. 119" (B. & W.).

Thompson, Edinb. New Phil. Jour., vol. xviii, p. 127, 1835; Ann. Mag. Nat. Hist., vol. xx, p. 157, 1847.

Templeton, Lond. Mag. Nat. Hist., vol. ix, p. 12, 1836.

Moore, Charlesworth's Mag. Nat. Hist., n. s., vol. ii, p. 206, 1838; ibid., vol. iii, pp. 196, 233, 1839.

Edwards, Annot de Lamarck, tom. v, p. 276, 1838; Hist. nat. des Crust., tom. iii, p. 145, 1840; Règne Anim. Crust., p. 197, pl. 67, f. 5, 1849.

Gould, Invert. Mass., pp. 338, 354, figure, 1840.

Fleming, Encyc. Brit., 7 ed., vol. vii, p. 502, 1842.

Dekay, Zool. New York, Crust., p. 48, pl. ix, fig. 33, 1844.

"Kirby and Spence, Int. Entom., 5th ed., p. 238; 6th ed., p. 203" (White.)

White, List Crust. Brit. Mus., p. 96, 1847; Brit. Crust. B. Mus., p. 68, 1850.

Dalyell, Powers of the Creator, vol. i, p. 241, pl. lxv, figs. 7-15, 1851.

Leidy, Jour. Acad. Nat. Sci. Phil., II, vol. iii, p. 150, 1855.

Gosse, Man. Mar. Zool., vol. i, p. 136, fig. 242, 1855.

Steenstrup and Lütken, Vidensk. Meddel., II, vol. ii, p. 275, 1861.

Hesse, Ann. Sci. nat., Zool., V, tome x. p. 113, 1868.

Jones, Trans. Nova Scotian Inst. Nat. Sci., vol. ii, pt. iv, p. 99, 1870.

Verrill, Proc. Am. Assoc., 1873, p. 367, 1874.

Macdonald, Trans. Linn. Soc., II, Zool., vol. i, p. 67, 1875.

Andrews, Q. Jour. Mic. Sci., II, vol. xv, p. 332, 1875.

Limnoria lignorum White, Pop. Hist. Brit. Crust., p. 227, pl. 12, fig. 5, 1857.

Bate, Rep. Brit. Assoc., 1860, p. 225, 1861.

Bate and Westwood, Brit. Sess. Crust., vol. ii, p. 351, figure, 1868.

Limnoria lignorum—Continued.

Norman, Rep. Brit. Assoc., 1868, p. 288, 1869.
Möbius, Wirbellos. Thiere der Ostsee, p. 122, 1873.
Parfitt, Fauna of Devon. Sess. Crust., p. (19), 1873.
Verrill, Am. Jour. Sci., III, vol. vii, pp. 133, 135, 1874; Proc. Am. Assoc., 1873, p. 371, 1874: This Report, part i, p. 379 (85), 1874.
Harger, This Report, part i, p. 574 (277), pl. vi, fig. 25, 1874; Proc. U. S. Nat. Mus., 1879, vol. ii, p. 161, 1879.
M'Intosh, Ann. Mag. Nat. Hist., IV, vol. xiv, p. 273, 1874.
Stebbing, Trans. Devon. Assoc., 1874, p. (8), 1874. Ann. Mag. Nat. Hist., IV, vol. xvii, p. 79, 1876.
Whiteaves, Further Deep-Sea Dredging, Gulf St. Lawrence, p. 15, "1874."
Metzger, Nordseefahrt der Pomm., p. 285, 1875.
Meinert, Crust. Isop. Amph. Dec. Daniæ, p. 77, 1877.
Smith, Proc. U. S. Nat. Mus., 1879, vol. ii, p. 232, fig. 2, 1880.
Limnoria uncinata Heller, Verh. k. k. Zool. bot. Ges. Wien, B. xvi, p. 734, 1866.
Stalio, Cat. Crost. Adriatic, p. 211, 1877.

PLATE IX, FIGS. 55–57.

This species may in general be recognized by its habits, being usually found burrowing in submerged timber, to which, notwithstanding its insignificant appearance, it often proves very destructive.

The body is subcylindrical, tapering slightly at each end and covered above with short hairs to which more or less dirt usually adheres. The head is narrower than the first thoracic segment. The eyes are lateral and consist of about eight ocelli, one central and the others around it. The antennulæ (pl. IX, fig. 56 a) are short and seem to arise from near the middle of the front of the head. The basal segment is the largest; the second and third are of slightly decreasing size; the fourth or flagellar segment is much the smallest, and tipped with setæ. The antennæ (pl. IX, fig. 56 b) are more slender than the antennulæ, and arise just below their bases and a little farther apart. The first two segments are short; the third slightly longer; the fourth and fifth increasing somewhat in length; the flagellum is not longer than the last two peduncular segments, and consists of a tapering segment, followed by a few short terminal segments provided with a terminal brush of setæ. The maxillipeds (pl. IX, fig. 56 c) are slender; the external lamella is semi-ovate, with the inner margin nearly straight, acute, and ciliated at the tip; the palpus is five-jointed but short, with the segments flattened, and all but the first ciliated along their inner margins. The outer maxillæ (pl. IX, fig. 56 d) are slender, three-lobed, and ciliated at the tip. The inner maxillæ (pl. IX, fig. 56 e) are also slender, the inner lobe tipped with pectinate bristles, the outer with robust spines. The mandibles (pl. IX, fig. 56 f) are somewhat elongate, but of a simple form, being curved inward, flattened and chisel-shaped at the tip; below there is a slight tubercle, apparently the rudiment of the molar process; externally, above the origin of the palpus, is a prominent tubercle; the palpus is short, of three subequal segments, the last furnished with a rather imperfect comb of setæ.

The first thoracic segment is about twice as long as any that follow; it is crossed by a broad, shallow depression, and is rounded at the sides.

The second and third segments are each about half the length of the first. The epimeral sutures are evident, and the epimera are rounded behind in the second segment, but a little more prominent in the third, becoming acute and increasing in size and extension backward to the seventh. The fourth segment is slightly shorter than the third, and perhaps a little broader; the last three are short, decreasing in length to the seventh, but maintaining about equal width. The legs are short and rather robust. The first pair have the carpus triangular, but this segment becomes more elongate in the succeeding pairs. The dactyli are robust, and are armed with a strong curved spine or claw at the tip and a smaller one below it. The merus, and usually the ischium and carpus, bear a few spiniform tubercles on the lower surface except in the last pair, which are also more elongated and slender than the others.

The pleon is scarcely narrower than the thorax, and tapers but little; the first four segments are of equal length; the fifth is longer with a median elevation and a transverse depression on each side. The last segment (pl. IX, fig. 57a) is transversely oval or subcircular, broader than long, with the anterior margin raised, especially at the middle, where the elevation is continued a short distance on the segment, but posteriorly it is flattened. The posterior margin is ciliate with hairs of various lengths. The uropods (pl. IX, fig. 57b) are attached just in front of the middle of the segment at its widest part. They consist on each side of a somewhat wedge-shaped basal segment, ciliated and bluntly denticulated distally on the outer side, and supporting two rami, between which it is produced below into a strong tooth-like process. The outer ramus is very short and curved outward; the inner is not as long as the basal segment, and is ciliated externally and at the tip. Underneath, the pleon is much excavated for the pleopods, which are strongly ciliated. The first pair (pl. IX, fig. 57c) consist on each side of a short basal segment bearing two lamellae; the inner lamella is almost four times as long as broad, with nearly parallel sides, ciliated at and near the tip; the outer, which is also in front of the inner, is sub-oval with the outer margin more convex than the inner, ciliated near the tip and along most of the outer margin, and inserted a little obliquely upon the basal segment. The next three pairs of pleopods are similar to the first pair on each side, except that in the males the second pair (pl. IX, fig. 57 d) bears a stylet (s) articulated to the inner margin of the inner lamella about the middle. The posterior pair of pleopods are smaller than the others and not ciliated.

Length 4.5^{mm}; breadth 1.5^{mm}; color light grayish.

Much has been written upon the destructive habits of the *Limnoria* or "gribble" and the means of preventing its attacks on woodwork, for which the reader may consult especially the publications of Leach, Coldstream, Hope, Thompson, Moore, Gould, Bate and Westwood, Verrill, and Andrews, who has observed it attacking the gutta-percha of submarine telegraph cables.

It is found boring in submerged wood along our coast from Florida! to Halifax!, N. S., and the Gulf of St. Lawrence. It occurs above low-water mark, but does not usually live far below that line; it has, however, been found by Professor Verrill at a depth of 10 fathoms in Casco Bay, and was dredged by the U. S. Fish Commission in a depth of 7½ fathoms, Cape Cod Bay!, Mass., in the summer of 1879. It is abundant, according to European authors, in many localities on the coast of Great Britain and in the North Sea. *L. uncinata* Heller, from Verbosca, in the Island of Lesina, Adriatic Sea, appears to be the same species, as the differences pointed out by Heller do not really exist, but were doubtless suggested by the incorrect figures that have been published representing the uropods with rami composed of two or more segments. The form of these appendages, as shown on plate IX, fig. 57 *b*, corresponds well with Heller's description. It was found by Heller associated with *Chelura terebrans*. *Limnoria* is said also to occur in the Pacific Ocean, and from its habits might be expected to have a wide distribution.

Specimens examined.

Number.	Locality.	Habitat.	When collected.	Received from—	Number of Specimens.	Dry. Alc.
2948	Florida	Boring in wood		Smithsonian Inst	6	Alc.
	Provincetown, Mass	do	Aug., 1879	U. S. Fish Com	90	Alc.
2917	Casco Bay	do	—, 1874	do	30	Alc.
	Bay of Fundy	do	—, 1872	do	60	Alc.
	Halifax, N. S	do	—, 1877	do	60	Alc.

IX.—CIROLANIDÆ.

Front formed of the approximate basal segments of the antennulæ, which are not covered by an anterior projection of the head; antennulæ and antennæ presenting an evident distinction into peduncular and flagellar segments; maxillipeds with a five-jointed palpus; mandibles formed for biting, palpigerous; legs all terminated by nearly straight dactyli; epimera distinct behind the first thoracic segment; pleopods at least the anterior pairs, ciliated; uropods biramous, the rami flattened and ciliated.

This family is represented on our coast by two closely allied species apparently belonging to the typical genus *Cirolana*, although approaching the allied genus *Conilera*, to which I formerly referred them. They have been hitherto usually referred to the following family, but the differences in the structure of the mouth parts, first pointed out by Schiödte, seem to warrant their separation as a distinct family. The mandibles are formed for biting, being armed with long and powerful teeth, which, closing together like the blades of scissors, are well adapted for lacerating the flesh of fishes on which they feed. The first three pairs of legs are fitted for prehension, but they are destitute of the strongly curved dactyli found in the *Ægida*, and still better developed in the *Cymothoidæ*. In the *Cirolanidæ* the propodus, in the first three pairs of legs, is somewhat curved and the dactyli are nearly straight, so that while the first three pairs of legs are powerful organs of prehension, they are also capable of letting go preparatory to the seizure of another victim. The posterior pairs of legs are ambulatory or fitted for swimming by their form and armature of bristly hairs. The ciliated pleopods are also powerful swimming organs, so that these animals are well fitted for the predatory life they lead. The epimera are well separated by sutures in all the thoracic segments behind the first. The pleon is scarcely narrower at base than the last thoracic segment, and is composed of six distinct segments, of which the last is much the longest, but not broader than the preceding segments, and tapers posteriorly. The uropods are lateral, articulated near the base of the last segment and distinctly biramous.

The mouth-organs of this and the two following families have been the object of special research by J. C. Schiödte, whose papers in the Naturhistorisk Tidsskrift have been in part translated in the Annals and Magazine of Natural History. He regards *Cirolana* as representing "the highest development of the crustacean type among the Isopoda," and even hints that *Cirolana* and *Ega* should be removed to opposite ends of the series of Isopoda. The same author would closely unite the *Bopyridæ*, *Ega*, and the *Cymothoidæ* into a single group, the *Cymothoæ*, while acknowledging that the young of *Cymothoa œstrum*, "according to the classification hitherto current, * * * would rather be allied to *Cirolana* than to *Cymothoa*." His classification, however, appears to be based almost entirely upon the structure of the mouth, disregarding the totality of structure upon which alone morphological classification can securely rest. In deference, however, to his views I have here regarded *Cirolana* as the type of a distinct family, which must still be considered as closely related with the two following families, on the principle that it is "more important that similarities should not be neglected than that differences should be overlooked."

Among the more important of the similarities by which these families seem to be united may be mentioned the following, as exemplified by our species. The segments of the thorax and pleon are all distinct from each other, so that the body, in the adults, appears to consist of thirteen segments behind the head, although in the genus *Ourozeuktes* Edwards* the segments of the pleon are consolidated. The epimera are distinct in all the segments behind the first thoracic. The pleon may or may not taper from the base, but it is terminated by a large scutiform segment, sometimes more or less sculptured, and bearing at the sides, near the base, a pair of uropods, in which the basal segment is more or less oblique distally and the rami lamelliform, though one of them may be narrowly so. The pleopods are unprotected by any form

---

* Hist. nat. des Crust., tome iii, p. 275, 1840.

of operculum and the anterior pairs are ciliated in the young of all three families, but this ciliation, as well as that on the uropods, may be lost in the sedentary adults of the *Cymothoidæ*. In all our species the dorsal surface is smooth throughout, or minutely punctate under a lens, but destitute of distinct roughness, tuberculation or sculpture, except that the telson may be faintly grooved or sculptured, and in some foreign species more distinctly so.

Cirolana Leach.

Cirolana Leach, Dict. des Sci. nat., tome xii, p. 347, 1818.

Thoracic segments subequal; eyes small, well separated; mandibles armed with strong acute teeth; dactyli straight, or but slightly curved; pleon of six distinct segments; basal segment of uropods with the inner angle produced.

Two closely allied species are found on this coast, which I formerly referred to the genus *Conilera* Leach. Further consideration induces me to refer them rather to the present genus, although they have some features which point toward *Conilera*, and are perhaps between that genus and the typical forms of *Cirolana*. From *Conilera*, as described by Bate and Westwood, our species differ principally in the more robust four posterior pairs of legs, in the produced angle of the basal segment of the uropods, and in the structure of the first pair of pleopods, which are not operculiform either in size or texture. Of these two species one is abundant and is described at length. The description will, however, apply almost equally well to the other except in the few points mentioned in the appropriate place. The characters given, though slight, appear to be constant, and I have therefore retained the two specific names.

This genus differs from *Ega* in the structure of the legs, and was placed by Professor Dana in a separate subfamily. In *Cirolana* the first three pairs of legs are strong, and armed with minute spine-like claws at the tip of the nearly straight dactyli; the propodi in these legs are robust, spiny, and somewhat curved, and some of the preceding segments are also armed with spines. These legs thus form powerful organs for seizing living prey, and are not, as in the *Cymothoidæ*, and, in a less degree, in *Ega*, merely fitted by their curved dactyli to retain the hold of the animal upon its host in a parasitic existence. The last four pairs of legs are well ciliated and capable of use either for walking or swimming, and these animals are thus fitted for their active and predaceous life.

Cirolana concharum Harger (Stimpson).

Ega concharum Stimpson, Mar. Inv. G. Manan, p. 42, 1853.
   Lütken, Vidensk. Meddel., 1859, p. 77, 1860.
Conilera concharum Harger, This Report, part i, p. 572 (278), 1874.
   Verrill, This Report, part i, p. 459 (165), 1874.
Cirolana concharum Harger, Proc. U. S. Nat. Mus., 1879, vol. ii, p. 161, 1879.

PLATES IX AND X, FIGS. 53–63.

This species may be most readily recognized among our Isopoda by the distinct thoracic and abdominal segments, the small lateral eyes, and the evident distinction, in both antennulae and antennae, of peduncle and flagellum. From the next species it is distinguished by the tip of the telson, which is truncated, or slightly emarginate, and grooved on the median line above near the end.

The body is, when extended, about three times as long as broad, and is smooth and polished throughout. The head is quadrate, a little broader in front than behind, and embraced at the sides by the first thoracic segment. The eyes are triangular, with the angles rounded, and are often partially covered below by the projecting anterior lobes of the first thoracic segment. They are separated by about three times their longest diameter. The antennulae (pl. X, fig. 60) are robust, with their basal segments in contact; the first segment is short and sub-spherical; the second also short; the third cylindrical and as long as the first two taken together and followed by a robust, but short, tapering flagellum, consisting of about fifteen segments, of which the second is as long as any other two, but the rest are all short. The flagellar segments beyond the first are provided each with a tuft of "olfactory setae." The antennae (pl. X, fig. 61 a) are longer and more slender than the antennulae, and are separated at their bases. The first four peduncular segments are robust; the first two short; the third and fourth each about twice as long as the first or second, and the fifth or last peduncular segment slightly the longest and much the most slender. The fourth and fifth segments bear along the distal portion of their outer margins long bristle-form hairs. The flagellum is slender and composed of from 15 to 18 segments, each bearing a few short bristles. The maxillipeds (pl. X, fig. 62 a) are elongated and almost pediform but flattened; the external lamella is small and subtriangular, rounded and hairy at the tip; the palpus is five-jointed, with the last four segments broad, flattened, and well ciliated; the tip of the maxilliped, nearly concealed by the large palpus, is provided with very densely plumose bristles. The outer maxillae (pl. X, fig. 61 b) are short and robust; the two articulated lobes narrow ovate, rounded at the tip, armed, especially the inner one, with spines and plumose or pectinated bristles. The inner maxillae (pl. X, fig. 61 c) are robust, with the outer lobe armed with strong smooth spines; the inner lobe rounded at the end and bearing three straight rather blunt spines, densely covered toward the tip with soft hairs. The mandibles (pl. X, figs. 61 d) are robust and horny at the tip, armed with one strong acute tooth, and in the right mandible with one acute and one obtuse tooth along a cutting edge, while the left mandible has three less acute teeth along this edge. Each mandible is, moreover, provided with a molar process or area (m), on its inner surface set along its interior and upper margin with spines. A narrowly lanceolate leaf-like appendage is attached just below the molar area. This appendage is furnished with a few bristles near the base, and its upper edge is armed with minute denticles; it is movable and ordinarily concealed behind the mandible. On the external surface, just above the origin of the palpus, each mandible bears two elevated, conical, obtuse tubercles. The palpi are slender, the second segment longest and hairy on the margin beyond the middle, the last segment slender and curved, with the usual hairs or slender bristles along the inner curvature.

The second and third thoracic segments are a little shorter than the others, which are of about equal length. The fourth and fifth segments are widest. The first segment is produced at the sides around the head so as to very nearly attain the anterior lateral angles of the head, and often so as to obscure the lower margin of the eyes. The epimeral sutures are scarcely distinguishable in this segment, but evident in the following segments. The epimera are rounded behind as far as the fourth, but the fifth is slightly angulated, and the sixth and seventh acute and produced backward beyond the margin of the corresponding segment. The first pair of legs are short and stout, and well armed with spines and bristles; the basis is of the ordinary form; the ischium is nearly triangular, having the upper margin much produced in the distal portion and bristly; the merus is expanded in a somewhat similar manner, but the angle is bent forward beyond the short carpus over the base of the propodus; the opposite or lower margin of the merus is armed with short stout spines; the carpus is short and small and possesses but little motion on the propodus, which is robust, somewhat curved, and bears a strong short dactylus. The second and third pairs of legs resemble the first pair, but the carpus increases somewhat in size, and there is more motion in its articulation with the propodus. They are directed forward, while the remaining pairs are usually directed backward and are more flattened. The fourth pair of legs are short like the first three (pl. X, fig. 62 b), but, except in size, resemble the following pairs. They are well provided with bristles in tufts, and along the margins of the segments, and especially the merus and two adjacent segments, are armed with long stout spines. The propodus is straight and much more slender than the carpus. The fifth and sixth pairs of legs increase in size, and the propodus especially becomes more elongated, but the seventh pair are a little smaller than the sixth.

The pleon is scarcely narrower at base than the last thoracic segment, and the first segment is often nearly concealed by the last thoracic. The fifth segment is longer on the back but shorter at the sides than the preceding segments. The last segment, or telson, is triangular with the ciliated apex truncated and emarginate or notched at the end of a short median furrow at the tip. The uropods (pl. X, fig. 63) slightly surpass the telson and are strongly ciliated; the inner ramus bears also a few spines near the tip; the basal segment has the inner angle produced along the margin of the inner ramus, which is broad and expanded distally, with a notch at the external angle; the outer ramus is slender and tapering, slightly surpassing the inner.

Length of large specimens 32^{mm}, breadth 10^{mm}, but usually smaller; 22^{mm} long, 7^{mm} broad. The ground color in life is yellowish, with reddish brown on the anterior margin of the head and on the posterior margins of the segments, especially in the dorsal region, where the segments are also marked with black dots. In life the body is somewhat translucent in the thinner parts. In alcohol the translucence disappears and the color fades to a nearly uniform yellowish or buff with black dots.

This species was described by Stimpson from Charleston, S. C. Most of the specimens in the collection are from Vineyard Sound!, where it occurs sometimes in great abundance, and is common especially during the winter. It is found swimming about in shallow water, and may be taken in a scoop-net, and is found also in lobster-pots. It was dredged in 45 fathoms off Block Island!, near the eastern end of Long Island Sound, in 1874, but has not yet been found north of Cape Cod.

Specimens examined.

Number.	Locality.	Fathoms.	Bottom.	When collected.	Received from—	Number of specimens.	Dry. Alc.
2061	Off Fishers Island			May —, 1875	J. H. Latham	100+	Alc.
2060	Vineyard Sound			Mar. —, 1874	V. N. Edwards	10	Alc.
2065	...do...	8.f.		Aug. 25, 1875	U. S. Fish Com.	1	Alc.
2062	Eel-pond, Wood's Holl		Muddy	July 23, 1875	....do......	100+	Alc.
2063	Off New Shoreham			Aug. 19, 1874	....do......	1	Alc.
2064	Off Martha's Vineyard	18	Sandy	Sept. 20, 1875	....do......	1	Alc.

Cirolana polita Harger (Stimpson.)

Ega polita Stimpson, Mar. Inv. Grand Manan, p. 41, 1853.
  Lütken, Vidensk. Meddel., 1859, p. 77, 1860.
  Verrill, Am. Jour. Sci., III, vol. v, p. 16, 1873.
Conilera polita Harger in Smith and Harger, Trans. Conn. Acad., vol. iii, pp. 3, 22, 1874.
  Verrill, Am. Jour. Sci., III, vol. vii, p. 411, 1874.
Cirolana polita Harger, Proc. U. S. Nat. Mus., 1879, vol. ii, p. 161, 1879.

This species so closely resembles the preceding, that a full description would be little else than a repetition of that given above. It appears, however, to differ constantly from the form already described, by its somewhat more elongated and cylindrical body; in the eyes, which are "elongate trapezoidal in shape, narrowest anteriorly," and in the tip of the telson, which is regularly rounded or slightly pointed at the tip without any truncation, much less any emargination, and is not at all grooved above.

Length 25^{mm}, breadth 6.5^{mm}; color much as in the preceding species.

Dr. Stimpson's specimens were "found on the fine sands at low-water mark on High Duck Island," in the Bay of Fundy, and the specimens that I have examined are from Cape Cod Bay!; from near Salem!, Mass.;

George's Banks!, and east of Banquereau!, or Quereau, latitude 40° 36′ north, longitude 57° 12′ west, where seven fine specimens were taken from a halibut (*Hippoglossus*), June 2, 1879, by Capt. J. W. Collins. It appears to replace the preceding species at the north.

Specimens examined.

Number.	Locality.	Fathoms.	Bottom.	When collected.	Received from—	Number of specimens.	Dry. Alc.
	Cape Cod Bay	7	Coarse, yellow sand.	Sept. 15, 1879	U. S. Fish Com.	2	Alc.
1314	George's Bank, lat. 41° 40′ N., lon. 6° 10′ W.	25	Sand	—, 1872	Smith and Harger.	1	Alc.
1399	George's Bank, lat. 42° 11′ N., lon. 67° 71′ W.	150	Soft, sandy mud.	—, 1872	Packard and Cooke.	1	Alc.
	Salem, Mass			—, 1878	J. H. Emerton.	1	Alc.
	East Quereau		190		June 2, 1879	Capt. J. W. Collins.	7	Alc.

X.—ÆGIDÆ.

Front formed of the approximate basal segments of the antennulæ, which are not covered by an anterior projection of the head; antennulæ and antennæ presenting an evident distinction into peduncular and flagellar segments; maxillipeds operculiform; mandibles formed for piercing, palpigerous, mouth suctorial; first three pairs of legs ancoral, last four ambulatory; epimera distinct behind the first thoracic segment; uropods lateral, biramous, ciliated, and flattened.

This family was represented within our limits by a single species of the typical genus until the summer of 1879, when a single specimen was collected of a second genus belonging to the *Ægidæ*, but having evident relations with the next family, and in many characters intermediate between *Ega* and the *Cymothoidæ*. The two genera by which the family is at present represented on our coast may be further characterized as follows: Both the antennulæ and the antennæ are directed laterally, the former arising near together on the anterior margin of the head and forming part of the outline of the animal as seen from above. They, as well as the antennæ, present an evident distinction into peduncular and flagellar segments. The maxillipeds are operculiform, and have the palpus armed with short hooks for adhesion to the surface of the fish on which they may be feeding. The mandibles are armed with a horny point, but not toothed as in the *Cirolanidæ*, and, while fitted for piercing, are not capable of lacerating and biting off pieces of flesh as in that family.

The first three pairs of legs are ancoral, or armed with strong curved dactyli, which, once implanted in the body of a victim, retain their hold without effort—a structure which attains its fullest development in the following family. The remaining pairs of legs are fitted for walking. The thoracic segments are subequal in length and have the epimera well separated, except in the first segment.

The pleon may or may not be suddenly narrower than the last thoracic segment, and, in our species, is composed of six distinct segments, of which the last is large and scutiform. The uropods are composed of a basal segment, oblique at the apex with the inner angle more or less produced, and bearing two flattened, ciliated rami; they are distinctly lateral, being inserted high up on the sides of the last segment.

This family contains our largest Isopod, *Ega psora*, and to it should probably be referred the huge *Bathynomus giganteus* A. Edwards, from the Gulf of Mexico, measuring more than eleven inches in length. It has usually been regarded as embracing the *Cirolanidæ*. I have already given my reasons for separating them, but have to regret my inability to examine many types of genera apparently more or less intermediate in position between *Ega* and, on the one hand *Cirolana*, and on the other *Cymothoa* and *Lironeca*. I have therefore retained the old classification rather than to unite the following genera with the *Cymothoidæ*.

Our two genera are most easily distinguished as follows: Eyes large and approximate, *Ega*, p. 89; eyes wanting, *Syscenus*, p. 93.

Æga Leach.

Ega Leach, Trans. Linn. Soc., vol. xi, p. 369, 1-15.

Eyes large; palpus of maxillipeds five-jointed; three anterior pairs of legs terminated by strong curved claws; posterior pairs slender, with slender nearly straight dactyli; pleon not suddenly narrower than the thorax; pleopods ciliated.

This genus is represented within our limits by a single species, which may be easily distinguished by its large approximate eyes. The basal segments of the antennulæ are flattened and the flagellum is comparatively slender. The maxillipeds have a five-jointed palpus, which is short and flattened and bent around the oral opening, and the inner margins of the three terminal segments are provided with a row of strong hooked spines, which are also found upon the outer maxillæ, thus forming two rows of short hooks on each side of the mouth, by means of which the opening of the mouth can be closely applied to the fish on which these animals prey. The inner maxillæ are slender and styliform and armed with sharp curved spines at the apex, and the mandibles are also acute and fitted for piercing. The body is moderately convex, and the last four pairs of legs are nearly alike ambulatory and of moderate length, the last pair, when extended, scarcely surpassing the telson. The pleon is composed of six distinct segments, and the basal segment of the uropods is strongly produced at its inner angle, as usual in the family. The pleopods are ciliated in the adults as well as in the young.

Æga psora Krøyer (Linné).

Oniscus psora "Linné, Fauna suecica, ed. ii, 1761"; Syst. Nat., ed. xii, tom. i, p. 1060, 1767.
   "Pennant, Brit. Zool., vol. iv, pl. 18, fig. 1, 1777 (certe)" (B. & W.).
   O. Fabricius, Fauna Groenlandica, p. 249, 1780.
   Mohr, Islandisk Naturhistorie, p. 110, 1786.

Æga emarginata Leach, Trans. Linn. Soc., vol. xi, p. 370, 1815; Dict. Sci. nat., tome xii, p. 349, 1818.
   Samouelle, Ent. Comp., p. 109, 1819.
   Desmarest, Consid. Crust., p. 305, pl. 47, figs. 4, 5, 1825.
   Griffith and Pidgeon, Nat. Hist. Crust., p. 218, pl. viii, fig. 3, 1833.
   Edwards, Hist. nat. des Crust., tome iii, p. 240, 1840; Règne Anim., Crust., pl. iv, fig. 4, and pl. lxvii, fig. 1, 1849.
   Gould, ?Rep. Geol. Mass., p. 549, 1835; Invert. Mass., p. 338, 1841.
   Gosse, Man. Mar. Zool., vol. i, p. 134, 1855.

Æga (*Oniscus psora*) Krøyer, Grönlands Antipoder, p. 318, 1838.

Æga psora Lilljeborg, Öfvers. Vet.-Acad. Förh., 1850, p. 84, and 1851, p. 21.
   Lütken, Vidensk. Meddel., 1858, pp. 65, 170, 1859; ibid., 1860, p. 181 **(7)** 1861; Crustacea of Greenland, p. 150, 1875.
   Schiödte, Ann. Mag. Nat. Hist., IV, vol. i, p. 12, 1868.
   Bate & Westwood, Brit. Sess. Crust., vol. ii, p. 283, figure, 1868.
   M. Sars, Chr. Vid. Selsk. Forh., 1868, p. 261, 1869.
   G. O. Sars, Hard. Fauna, Crust., p. 275 [32], 1872.
   Verrill, Am. Jour. Sci., III, vol. v, p. 16, 1873.
   Smith and Harger, Trans. Conn. Acad., vol. iii, p. 22, 1874.
   Whiteaves, Further Deep-Sea Dredging, Gulf St. Lawrence, p. 15, "1874."
   Metzger, Nordseefahrt der Pomm., p. 285, 1875.
   Meinert, Crust. Isop. Amph. Dec. Daniæ, p. 89, "1877."
   Miers, Ann. Mag. Nat. Hist., IV, vol. xix, p. 134, 1877.
   Harger, Proc. U. S. Nat. Mus., 1879, vol. ii, p. 161, 1879.

Æga entaillé Latreille, Règne Anim., tome iv, p. 134, 1829.

PLATE X, FIG. 64.

The present species is the largest Isopod, and indeed the largest Tetradecapod known on the New England coast, reaching a length of nearly or quite two inches and a breadth of one inch, and has even attained to the dignity of a popular name, "salve-bug", by which it is known among fishermen. It may be further distinguished by its large approximate eyes, covering a large proportion of the upper surface of the head, and by the possession of ancoral legs in three pairs only, the last four pairs of legs being fitted for walking.

The body is oval, broadest at the fourth and fifth thoracic segments, where the breadth is about half the length. The dorsal surface is moderately convex and smooth except for minute and rather scattered punctations, which occur also on the legs, especially on the basal segments, on the antennulæ, the uropods, and even the pleopods. The head is transverse and sub-triangular, salient in front between the bases of the antennulæ. Much of the upper surface of the head is covered by the large oval or somewhat reniform eyes, which do not quite meet on the median line. The antennulæ when bent backward nearly or quite attain the anterior margin of the first thoracic segment, and have their first two segments large and flattened, and wedge-shaped in front; of these the basal segment is quadrate in outline, as seen from above, and nearly as broad as long; it closely approaches its fellow of the opposite side in front, but is separated from it behind by a median process of the head; the second segment is triangular in outline, as seen from above, with the apex of the triangle extending beyond the origin of the third slender cylindrical segment, which is followed by a tapering flagellum of about a dozen segments. The antennæ when reflexed extend beyond the first thoracic segment and have the first two segments short and compressed, the third somewhat longer, the fourth and fifth longer and nearly cylindrical, followed by a tapering flagellum about as long as the peduncle and composed of fifteen to twenty segments. The maxillipeds have a short triangular external lamella and a five-jointed palpus, of which the first segment is short and transverse; the second is triangular and bears, on its inner apex, a few slender hooked spines; the third segment is broad and flattened, with the inner margin short, and armed with about three robust hooked spines; the fourth segment is flattened and transverse and armed along its inner margin with about six similar spines; while the fifth segment is small, sub-oval, and armed with much more slender curved spines. The outer maxillæ are provided with curved spines at the apex much like those of the maxillipeds. The inner maxillæ are rod-like and terminate in sharp somewhat curved spines placed close together. The mandibles support a slender palpus of three segments, of which the middle one is much the longest, and the last is robust and sickle-shaped, with a comb of short spines along the inner curve. This segment lies, in the ordinary position, just at the base of the antenna of the same side.

The first thoracic segment is, at its anterior margin, scarcely broader than the head, but expands rapidly backward. It is excavated in front for the eyes, which project somewhat beyond the posterior margin of the head. The second, third, and fourth thoracic segments are each a little shorter than the first; the fifth and sixth are somewhat longer; the seventh is shorter than the sixth. The epimera of the first thoracic segment are not separated by suture, but in the second and following segments they are so separated, and, especially on the anterior segments, marked with two oblique depressed lines. The epimera of the second, third, and fourth segments are rounded or truncate behind, but in the posterior segments they become acute and extend beyond the angles of the segments to which they are attached. The first three pairs of legs are short and armed with strong hooked dactyli. The propodal segments are also curved, and the carpus is short in the first pair but somewhat longer in the second and third pairs. The merus is almost crescent-shaped in the first pair of legs, its horns embracing the carpus above and below, but it becomes more elongated in the succeeding pairs; in all three pairs its inferior margin is armed with a few short, stout spines. The fourth and succeeding pairs of legs are of quite a different type from the first three. The four segments following the first or basal one are straight, cylindrical, or slightly compressed, armed with short spines, especially below and at the distal end, subequal in length but decreasing in diameter to the propodus, which bears in each pair a short, slightly curved and comparatively weak dactylus. The seventh pair is only imperfectly developed in the young specimen figured, but never quite attains the size of the sixth pair, which is the largest.

The pleon is scarcely narrower than the last thoracic segment and tapers but little to the fifth segment. The last segment is triangular, with the sides but little dilated, and is pointed at the tip without grooves or carinations. The uropods scarcely surpass the telson; the basal segment has its inner angle long and spiniform, extending the whole length of the inner margin of the inner ramus and ciliated toward the tip; the rami are flattened, the outer elongate ovate, obtuse; the inner with the inner margin straight, the outer curved and emarginate near the tip. Both rami and the posterior part of the telson are ciliated.

Length 16–50^{mm}, breadth 7–25^{mm}; color in alcohol light brown, darker toward the head; eyes black.

Linné's description of *Oniscus psora* is too indefinite to be certainly recognizable, and in using his trivial name I have followed the authority of Lütken and others. Our specimens agree well with the description of *O. psora* by O. Fabricius, and are undoubtedly identical with that species, which he describes as infesting the cod. They appear to correspond also with Bate and Westwood's figure and descriptions, although those authors make no mention of Fabricius under *Æ. psora*. As Kröyer referred the species to its proper genus, I have adopted his name as authority for the combination.

The specimen figured was dredged in the summer of 1872, a little to the northeast of St. George's Bank!, in latitude 42° 11' north, longitude 67° 17' west, in 150 fathoms, soft sandy mud with a few pebbles, and is young, as shown by its size and imperfectly developed seventh pair of legs. Adults may surpass the size of the figure, but the specimen drawn was enlarged three diameters. Adult specimens were obtained from the Provincial Museum, Halifax, Nova Scotia, labeled as found on the cod, and were probably from the fishing banks of that region, or from the Banks of Newfoundland. During the summer of 1879 a considerable number of specimens were received by the Fish Commission through the Gloucester fisheries, of which only a few are included in the table of specimens examined. These specimens were parasitic on the cod (*Gadus morrhua*), and on the halibut (*Hippoglossus*). Specimens have also been obtained from the skate (*Raia*). Whiteaves records this species from a halibut, on the north shore of the Gulf of St. Lawrence. Fine specimens were obtained by Mr. N. P. Scudder from off Holsteinborg, Greenland, in Davis' Straits!, parasitic on the halibut, and collected in July and August, 1879. It extends to Iceland (Edw. *et al.*); the British Isles (B. and W.); the North Sea (Metzger); Finmark (Sars), and Spitzbergen (Miers).

Specimens examined.

Number.	Locality.	Fathoms.	Parasitic on—	When collected.	Received from—	Number of specimens.	Dry. Alc.
1398	George's Bank, lat. 42° 11' N., lon. 67° 17' W.	150		—, 1872	Packard and Cooke	1	Alc.
2139					Colonial Mus., Halifax.	2	Alc.
	George's Bank			—, 1878	Schooner Alice G. Wonson.	3	Alc.
	....do		Codfish	May 8, 1879	J. P. Shomelia	3	Alc.
	....do		....do	May 15, 1879	Capt. J. Q. Getchell	9	Alc.
	N. E. George's Bank	47	....do	Nov. 29, 1878	J. P. Shemelia	3	Alc.
2151	Gulf of Maine		Skate (*Raia*)	—, 1878	U. S. Fish Com'n	20	Alc.
2156	Banquereau		Halibut	—, 1878	Schooner Marion	1	Alc.
2157		40–50	Codfish	—, 1878	Schooner Rebecca Bartlett.	1	Alc.
2158	Grand Menan Bank	100		—, 1878	Schooner Peter D. Smith.	3	Alc.
2155	....do	100		—, 1878	U. S. Fish Com'n	1	Alc.
	Brown's Bank	52	Codfish	Dec. 19, 1878	Mr. Isaac Butler	2	Alc.
	....do		....do	Feb. 13, 1879	Capt. J. Q. Getchell	2	Alc.
	....do	30	....do	May 1, 1879	....do	8	Alc.
	Lat. 43° 25' N., Lon. 60° W.	180	Halibut	Aug. 21, 1879	Capt. S. W. Smith and crew.		Alc.
	Davis's Straits		....do	—, 1879	Mr. N. P. Scudder	10	Alc.

Syscenus * gen. nov.

Eyes wanting; palpus of maxillipeds two-jointed; sixth and seventh pairs of legs elongated; pleon suddenly narrower than the thorax; pleopods naked.

This genus is unfortunately represented in the collection by a single specimen. It differs from *Ægа* by characters that point toward the *Cymothoidæ*, as in the reduction of the segments of the palpus of the maxillipeds, the sudden constriction at the base of the pleon, and the naked pleopods. The absence of eyes, although a conspicuous character can hardly be regarded as of great taxonomic value. It is separated from the *Cymothoidæ* by the form of the head, which is not produced over the bases of the antennulæ but merely projects slightly between them. The antennulæ moreover are composed of three peduncular segments and a flagellum; the basal segments are much smaller than in *Ægа* and less flattened, but still form a part of the anterior outline when seen vertically. The last four pairs of legs differ from the first three, and are more or less elongated and fitted for crawling. The uropods are distinctly ciliated.

Syscenus infelix sp. nov.

This species may be recognized among our Isopoda by the possession of the full number of segments, the ciliated uropods, naked pleopods, and the absence of eyes.

---
* Σύσκηνος, a messmate.

The body is more than twice as long as broad and only moderately convex. The head is small and as seen from above is transversely somewhat diamond-shaped with rounded angles. It presents in front a slight prolongation between the antennulæ, and on each side of the short median process its outline is excavated above the bases of the antennulæ. The posterior margin is curved, but near each end is a faint indication of a lobe, projecting backward like the ocular lobes in *Ega*, but the eyes are wanting. The antennulæ arise near together on each side of the front and are short, extending when reflexed but little beyond the lateral margins of the head and only slightly surpassing the fourth antennal segment. They are readily distinguishable into peduncular and flagellar segments, the first three segments being of comparatively large size and about equal length; the second segment much flattened below against the antennæ; third more slender than the first two and followed by a short, tapering six-jointed flagellum. The antennulæ are in their natural position reflexed, the second segment being articulated at an angle with the first. The antennæ are considerably longer than the antennulæ and, when reflexed, slightly surpass the posterior border of the third thoracic segment. They are inserted below and a little outside of the antennulæ. The first segment is short and flattened below; the second is also short, the two together being hardly longer than the basal antennular segment; the third segment is about as long as the first two together, and the fourth is a little longer than the third, but of slightly less diameter; the fifth is more than one-half longer than the fourth, but is more slender and is followed by a slender, tapering flagellum of about twenty-four segments. The last two peduncular segments bear a row of elongate bristly hairs along the margin which, when reflexed, is brought next the body, and the row is continued, though with shorter hairs, along the flagellum. The palpus of the maxillipeds is composed of two segments of which the first is nearly square and armed at the inner distal angle with a minute hook; the second is bluntly triangular and armed at the apex, which is directed inward, with three hooklets. The external lamella is small and subcircular. The outer maxillæ are armed with short hooks at the tip; the inner with minute denticles. The mandibles are flattened and denticulate at the tip and bear a three-jointed palpus of which the three segments decrease in size to the last.

The first thoracic segment is twice as long as the second; its anterior margin is adapted to the head; its posterior margin is nearly straight above and rounded at the sides until the epimeral region is reached, when a short, pointed projection juts backward, being the tip of the epimeron on each side, here united with the segment. The next three —second, third and fourth—thoracic segments are of about equal length, and each a little over half the length of the first segment; their posterior margins are nearly straight above and rounded at the sides; the third segment is broadest. The fifth and sixth segments are each a little longer than the second; the seventh about as long as the second. The last segment, and in a less degree the sixth and fifth segments, have their posterior margins excavated along the back; all have their lateral angles rounded, although the angles of the seventh segment are but slightly so. The epimera are short and pointed; those belonging to the second and third segments are larger than the following ones, and are applied directly to the lateral margin of the segments; the posterior four pairs of epimera are shorter and smaller, and are separated from the lateral borders of the segment by a fold of the integument cutting off a portion of the anterior lateral angle and increasing in size to the last segment.

The first three pairs of legs are alike, distinctly ancoral and directed forward. In each the basis is much the longest segment; the ischium is strongly flexed upon it; the merus is expanded distally around the base of the carpus and bears a few bristles at the outer angle; the carpus is short, less than half as long as the propodus, and the dactylus is strong and curved. The fourth pair of legs, like those that follow, is directed backward; the basis is the longest segment and the ischium is strongly flexed upon it and of more than half its length; the merus, carpus and propodus are each about two-thirds as long as the ischium, and all four segments are armed distally with a whorl of spines around the articulation with the succeeding segment; the dactylus is slender, sharp and curved. The fifth pair of legs is longer than the fourth by a little more than the length of the dactylus, the elongation being in the segments from the ischium to the propodus inclusive. The sixth pair is the longest, being, when extended, as long as the thorax and pleon together. This elongation is confined also to the four segments above indicated, and of these the ischium is about as long as the basis; the merus falls a little short of the ischium in length; the carpus and propodus are of equal length, and are as long as the ischium; all these segments are slender and slightly curved, and are armed distally and along their inner side with short spinules. The dactylus is slender and curved. The seventh pair of legs resembles the sixth but is shorter by about half the length of the propodus. The fifth pair does not attain the middle of the carpus of the sixth.

The pleon is of less diameter than the last thoracic segment and about as long as the last five thoracic segments. Its transverse diameter increases slightly to the base of the last segment, where it is broadest; the fifth segment is a little longer than the preceding one, and the last segment is of a broad ovate form, acuminate and ciliated at the tip, truncated at the base and smooth above, except for a faint transverse impression on each side near the base, and a still more faint impressed median line toward the tip. The uropods attain the tip of the telson but do not surpass it; they have the basal segment oblique but not produced at the inner angle, and bearing two elongate-elliptical rami, tapering at the base and ciliated, the inner about one-third longer than the outer. The pleopods are quite naked and destitute of cilia.

Length 25^{mm}; breadth, 9^{mm}; breadth of pleon 4^{mm}; length of head 3^{mm}; breadth 4.2^{mm}.

A single specimen of this species was dredged by the U. S. Fish Commission, about fifteen miles northeast of Cape Cod!, in 150 fathoms brown mud, September 10, 1879.

XL.—CYMOTHOIDÆ.

Head produced anteriorly over the bases of the antennulæ; maxillipeds few jointed, operculiform; mandibles palpigerous; mouth suctorial; legs armed with strong curved dactyli; epimera distinct behind the first thoracic segment; telson large and flattened; pleopods not ciliated; uropods articulated near the antero-lateral angles of the last segment, and composed of a more or less flattened basal segment bearing two flattened rami; habit parasitic; body often unsymmetrical by distortion in the adults.

This family is represented within our limits by three genera and as many species. They are parasitic in habit, usually on fish, and fix themselves by their strongly curved claws to their host, often within the mouth, or about the branchial cavity, and frequently become distorted when fully grown. In all our species the head is small, and has the anterior margin produced, concealing the bases of the antennulæ and the antennæ. The head is three-lobed behind, and the first thoracic segment is adapted to it. The antennulæ and antennæ are both short and tapering, without very evident distinction into peduncular and flagellar segments. This distinction is, however, usually more or less evident on examination.

The epimera are well separated, except in the first segment, and may be projecting and conspicuous. The legs are of nearly the same form throughout, but increase in length and become more slender posteriorly.* The basal segments are in some genera enlarged and flattened, but not in ours; the joint between the basis and ischium is strongly flexed, and the segments, at least beyond the ischium, to the dactylus, are short and capable of but little motion on each other. The dactylus is strongly curved and admirably fitted for firm attachment to the host on which the animal may be living. In our species the legs, in the natural position, are concealed in a dorsal view beneath the body of the animal, to the under surface of which they are appressed, the first three pairs being directed forward, and the last three backward, as represented in plate X, fig. 66.

The pleon in our species is not suddenly narrower than the thorax, as it is, however, at least in the adults, in some genera belonging to this family. The segments of the pleon are distinct, the last one scutiform

---

* In *Ichthyoxenus Schmidtii* the seventh pair of legs "retach to the extremity of the first and are slender, compressed crawling legs, with a small, almost rudimentary, straight claw."

and of moderate size, not being greatly enlarged. The pleopods are destitute of cilia in the adults.

This family is evidently closely related to the preceding and may yet have to be united with it, or even be extended so as to include also the *Cirolanidæ*. Our representatives of the three families are so few that I have had little opportunity to study the genera, and as before stated, I have separated the *Cirolanidæ* principally in deference to the opinions of Schiödte. *Alitropus* Edwards, *Syscenus* Harger, and *Egathoa* Dana may be mentioned as genera pointing toward a transition between the *Ægidæ* and *Cymothoidæ*, and it is evident that the latter family is made up of forms degraded by parasitism. They have thus exchanged the ambulatory legs of the *Ægidæ* for strictly ancoral legs, for the most part in seven pairs, and have lost the natatory cilia of the pleopods. Their antennary organs are also much less perfect than in that family. All these modifications are in the line of the sedentary life of a parasite.

The interesting observations of Mr. J. F. Bullar have shown that in certain genera of the *Cymothoidæ* (*Cymothoa*, *Nerocila*, *Anilocra*) a peculiar form of hermaphroditism occurs, the young at a certain stage of development being males with well developed testes and external organs, but possessing at the same time ovaries with the oviduct ending blindly. As development proceeds the male organs are lost by molting, the oviduct obtains an external opening, the incubatory pouch is developed, and the animal becomes a female. Mr. Bullar's statements provoked considerable discussion, but they have recently been verified by Mayer, who has, however, shown that self fertilization does not occur.

Three genera of *Cymothoidæ* are represented within our limits by as many species, and a fourth species, *Cymothoa prægustator* Say* (Latrobe) may yet be found, being not a rare parasite in the mouth of the menhaden (*Brevoortia menhaden* Gill) in southern waters. The projection of the front of the head over the bases of the antennary organs, and the strongly hooked or ancoral legs are characteristic of the family, and the genera may be distinguished by means of the following table:

	ciliated; eyes large conspicuous,		*Egathoa*, p. 393
Propods	naked; body	symmetrical; posterior epimera elongated,	*Nerocila*, p. 394
		asymmetrical; epimera short,	*Livoneca*, p. 394

Nerocila Leach.

Nerocila Leach, Dict. Sci. nat., tom. xii, p. 351, 1818.

Body oval; head small; eyes of moderate size; posterior thoracic segments and epimera angulated or spiniform, giving a sharply serrated or dentated outline to the thorax; first two "abdominal epimera" also spiniform; pleon of six distinct segments.

Our species of *Nerocila* has the characters of the genus much less pronounced than some foreign ones, as the posterior epimera are nearly

*Jour. Acad. Nat. Sci. Phila., vol. i, p. 395, 1818.

or quite concealed from above by the projecting angles of the segments, and the "abdominal epimera" are mostly concealed beneath the pleon. These organs are the much elongated inferior angles of the segments, which in allied genera, as *Ægathoa*, are short and not produced. In a lateral view they considerably resemble the posterior epimera, giving the appearance of two additional pairs. The specimen first described is smaller than others that have since been obtained.

Nerocila munda Harger.

Nerocila munda Harger, This Report, part 1, p. 571 (277), 1874; Proc. U. S. Nat. Mus., 1879, vol. ii, p. 161, 1879.

Verrill, This Report, part i, p. 459 (165), 1874.

PLATE X, FIG. 65.

This species may be recognized among our Isopoda by the projecting posterior epimera, and the two pairs of spiniform "abdominal epimera" beneath the pleon.

The body is oval, twice as long as broad, smooth, polished, and moderately convex. The head is flattened, broader than long, narrowing anteriorly, broadly rounded or subtruncate in front, three-lobed behind, with the middle lobe largest. The eyes are black and consist of an irregularly rounded patch of small indistinct ocelli, and are visible both above and below. The antennulæ are about as long as the head, and composed of eight segments, of which the first is short, the second is the longest, and the remaining six decrease pretty regularly in size to the last. The antennæ are a little longer and more slender than the antennulæ and have the first segment short, the second subglobose, the third, fourth, and fifth cylindrical, and a little larger than the segments of the flagellum, which are about five in number. The mandibular palpi are longer than any three segments of the antennæ, and the first segment is large, the second elongate conical, the third shorter, cylindrical.

The first thoracic segment is much longer than the succeeding ones and adapted to the head in front. It is slightly produced at its lateral angles behind, or rather appears so from the union of the epimera, which really constitute the projecting angles to the segment. In the second, third, and fourth segments the posterior angles are but little produced, and are equaled or slightly surpassed by the epimera, but in the last three segments the posterior angles are acutely produced much beyond the epimera of the corresponding segments, the angle of the sixth segment nearly attaining the end of the seventh epimeron. In a lateral view, only the last two epimera are decidedly acute, while those of the second and third segments are obtuse and rounded behind. Seen from below, the posterior angles of the epimera are acute throughout. The first pair of legs are slightly more robust than the second and third; the last four pairs are still more slender, the last pair longest, and the last two pairs armed with a few short spinules.

The pleon is shorter than the thorax and much narrower, though not suddenly so and tapers but little posteriorly; the telson is flattened, and regularly rounded behind. The "abdominal epimera" are acute, the second smaller and more slender than the first, but their extension backward varies with the state of contraction of the pleon. The uropods (pl. X, fig. 65 a) surpass the telson, and have the inner angle of the basal segment sharply produced. The rami are flattened; the external one twice the length of the basal segment, narrowly ovate or lanceolate, sometimes slightly curved, and surpassing the telson by half its length. The inner ramus is narrowly oval, obliquely truncate behind and about three-fourths as long as the outer.

The length of the specimen figured, which was the one first described, is 15^{mm}, breadth 7^{mm}, but specimens measuring 25^{mm} in length have since been collected; color brown or greenish, with two narrow dorsal bands of lighter color, most evident at the extremities.

The original specimen was obtained on the dorsal fin of *Ceratocanthus aurantiacus* at Wood's Holl!, Vineyard Sound, in 1871, and two more specimens of larger size have since been obtained, also from Vineyard Sound!, Mass.

Ægathoa Dana.

Ægathoa Dana, Am. Jour. Sci., II, vol. xiv, p. 304, 1852.

Body elongate oval; pleon not suddenly narrower than the thorax; head large, subtriangular; eyes large; legs nearly alike throughout, with strong curved dactyli; epimera of moderate size or small; pleon long and large, composed of six distinct segments; pleopods not ciliated; uropods more or less distinctly ciliated, rami subequal.

This genus is represented in our fauna by a species parasitic in the mouth of a squid. The large, granulated eyes remind one of *Æga*, and the ciliated uropods also indicate the approximation of this genus to the preceding family. The ciliation is, however, nearly rudimentary in our species, and is present, at least in the young, of other members of the *Cymothoidæ*.

Ægathoa loliginea Harger.

Ægathoa loliginea Harger, Am. Jour. Sci., III, vol. xv, p. 376, 1878; Proc. U. S. Nat. Mus., 1879, vol. ii, p. 161, 1879.

PLATE X, FIG. 66.

The legs all armed with strong curved claws, the large conspicuous eyes and the slightly ciliated uropods serve to distinguish the present species from the other Isopoda of our coast.

Body elongate oval in outline, nearly four times as long as broad, slightly dilated near the posterior end. Head broadly rounded in front, subequally, but not deeply, trilobed behind. Eyes large, with evident facets, lateral, semi-hexagonal, visible from below, covering nearly half the area of the head above, projecting posteriorly beyond the middle lobe of the head. Exteriorly they form about two-thirds of the lateral margin of the head. Their interior boundary is in the form of three sides of a hexagon, separated at their nearest points by a little more than the transverse diameter of the eye. The antennulæ are about as long as the head, composed of eight segments and separated at the base. The first segment is short and stout; the next two a little longer, but scarcely distinguishable from the following five flagellar segments, which decrease in size to the last. The antennæ are composed of ten segments. They are more slender than the antennulæ, and surpass them by about two segments. The first two segments are broader than the following three, which are also somewhat larger than the five flagellar segments.

The first thoracic segment is shorter than the head, but much longer than any of the succeeding segments, which to the sixth are of equal length, each about one-third shorter than the first. The seventh segment is about one-third shorter than the sixth. The fifth and sixth are broadest, each being about one-third broader than the first. The epimera do not project behind the angles of the segments to which they are attached. The legs differ but little throughout. The first pair are shortest, and the first three pairs are somewhat stronger than the last four, which are armed with a few scattered short spinules. The seventh pair are the longest.

The pleon is a little longer than the seven thoracic segments. The fifth segment is broader behind than in front, and the last segment is as broad at the insertion of the uropods as the third segment, and is rounded behind. Anterior pleopods with the basal segment nearly square. The uropods are unlike on the opposite sides in the specimen figured. The normal form is probably seen in the right uropod, which surpasses the telson by less than half the length of the outer ramus. This ramus is longer than the inner, narrow, with nearly parallel sides and is obliquely truncated at the tip. The inner ramus is somewhat diamond-shaped. The ciliation is nearly rudimentary and might be overlooked. The basal segment is alike on the two sides and has the inner distal angle acute and but slightly produced.

Length 13^{mm}, breadth 3.6^{mm}; color in alcohol yellowish, with minute black specks most abundant on the pleon; eyes black, conspicuous.

The specimen was obtained June 1, 1874, by Mr. S. F. Clark, at Savin Rock!, near New Haven, from the mouth of a squid (*Loligo Pealii*), whence the specific name. Two specimens "parasitic on young mullet" are in the Yale College Museum, collected at Fort Macon!, N. C., by Dr. H. C. Yarrow, which appear to belong to this species, showing that it is not confined to the squid.

Livoneca Leach.

Livoneca Leach, Dict. Sci. nat., tome xii. p. 351, 1-15.

Head small, projecting in front over the bases of the antennulæ, which, like the antennæ, are short; legs all alike and armed with strong curved dactyli; body broad, oval, often obliquely distorted.

This genus is represented by a single species, in which the body is of a broadly oval form and depressed. All the legs are short and armed with strongly curved dactyli, and, in the natural position, are closely appressed to the ventral surface, which, however, is more or less exposed below along the middle.

Livoneca ovalis White (Say).

    Cymothoa ovalis Say, Jour. Acad. Nat. Sci. Phil., vol. i, p. 394, 1818.
    Dekay, Zool. New York, Crust., p. 48, 1844.
    Livoneca ovalis White, Cat. Crust. Brit. Mus., p. 109, 1847. (*Livoneca*).
    Harger, This Report, part i, p. 572 (278), pl. vi, fig. 29, 1874; Proc. U. S. Nat. Mus., 1879, vol. ii, p. 162, 1879.

    PLATE XI, FIG. 67.

The broadly oval, more or less distorted and unsymmetrical form of this Isopod serves to distinguish it from any other species yet recognized within our limits.

Body broad, oval, usually oblique, and not, as represented in part I of this report, pl. VI, fig. 29, with the sides of equal length. The legs, moreover, in that figure are in an unnatural position, as they are, during life, concealed beneath the body of the animal and appressed to the ventral surface, the first three pairs directed forwards and the last four pairs backward. The dorsal surface is moderately convex. The head is small, rounded in front, trilobed behind, the middle lobe much the largest, the two lateral lobes extending beyond the eyes, which are not conspicuous, small and broadly separated. Antennulae (pl. XI, fig. 67a) widely separated at the base, with the first segment short and stout; the second longer and somewhat tapering; the third about as long as the first. These peduncular segments are somewhat flattened. The flagellum is longer than the peduncle, tapering and five-jointed, curved backward in the natural position, each segment bearing a row of short blunt setae, near the distal end, on the inner curve. The antennae (pl. XI, fig. 67b) are about as long as the antennulae, with the first two segments short and stout, the next three more slender; flagellum three or four jointed, with the last segment imperfectly divided and tipped with a few short setae. The maxillipeds are narrow, with the outer lamella partially united to the basal segment and the palpus tapering and two-jointed, tipped with a few short curved setae, at least in young individuals. The mandibles are pointed; their palpi (pl. XI, fig. 67c) tapering from the base and composed of three segments of about equal length, the first subquadrate, the second tapering, the third nearly cylindrical.

The first thoracic segment is longest; the next three a little shorter and about equal; the fifth and sixth still shorter; the seventh shortest, measured along the median line, which is usually a curved line except in young specimens. The anterior margin of the first thoracic segment is adapted to the posterior margin of the head and presents three sinuses, the middle one largest, for the median lobe of the head, and two smaller ones for the ocular lobes. The posterior margin of this segment is strongly convex backward throughout. In the succeeding segments this convexity rapidly diminishes so that the fourth has nearly a transverse margin and the last three segments become concave behind in an increasing degree. The epimera are narrow and obtusely pointed behind, and do not surpass the posterior angle of the segment to which they are attached except in the last two segments. The first pair of legs (pl. XI, fig. 67 d) are short and stout, the basal segment large but short; the next three segments short and with little motion on each other; the propodus stout and somewhat curved; the dactylus long, curved, and strong. The second and third pair of legs are much like the first, as are the four succeeding pairs, but somewhat larger and longer. The seventh pair (pl. XI, fig. 67 e) have the basal segment about twice as long as in the first pair, and the succeeding segments are also proportionally longer than in the first pair, except the dactylus, which is slightly weaker and not longer than in the first pair.

The pleon tapers rapidly at the sides; its first five segments are subequal in length; the last segment forms about half its length, and is flat and broadly rounded behind. Uropods (pl. XI, fig. 67 f) surpassing the telson with the basal segment, about as long as the rami and but little produced at its inner angle; outer ramus linear oblong, rounded at the end; inner ramus shorter and broader, oblique at the tip.

Length 17–22^{mm}, breadth 10–12^{mm}. These animals when preserved in alcohol are of a leaden color, with the posterior margins lighter.

They are often parasitic on the blue-fish (*Pomatomus saltatrix* Gill). The details figured on plate XI are from small specimens collected on young blue-fish at New Haven!, by Mr. F. S. Smith. Other localities are Thimble Islands!, Long Island Sound; Vineyard Sound!, Fish Commission 1871, one specimen among scup (*Stenotomus argyrops* Gill). A specimen was sent to the Museum in 1878, collected by Dr. T. H. Bean, from the gill of *Micropogon undulatus* caught at Norfolk!, Va., July 9, 1878.

Specimens examined.

Number	Locality.	Parasitic on—	When collected.	Received from—	No. of specimens.	Dry. Alc.
....	Norfolk, Va	Micropogon	July 9, 1878	T. H. Bean	1	Alc.
2071	New Haven	Blue-fish	....	F. S. Smith	15	Alc.
2072	Vineyard Sound	do	——— —, 1871	U. S. FishCom	1	Alc.
2073	....do	Scup	Aug. 17, 1871	...do	1	Alc.
2074						Alc.
2075	Vineyard Sound	Blue-fish	Sept " 1871	U S Fish'Com	1	Alc.
				F. H. Bradley	1	Alc.

XII.—ANTHURIDÆ.

Body elongate, cylindrical; mouth suctorial; legs ambulatory and prehensile, the first pair enlarged; first pair of pleopods thickened and crustaceous, protecting the following pairs; uropods articulated at the sides of the last segment, standing in a more or less vertical position and forming with the telson a sort of cup or flower at the end of the body.

This family is represented within our limits by three species belonging to as many genera, which, in addition to the characters given above, agree further in the following particulars: The body is elongated and vermiform, often more than ten times as long as broad, and of nearly uniform size throughout. The head and thoracic segments are all distinctly separated from each other, and the head and last thoracic segment are shorter than the intervening segments, which are subequal. Both pairs of antennæ are approximate at their bases, and the lower pair or true antennæ are short, not greatly surpassing the head in length. These organs have the basal segment short, the second segment flattened internally and adapted to its fellow of the opposite side, while above and externally it is excavated for the basal segment of the antennulæ. The mandibles are palpigerous, and the mouth parts are fitted for piercing and for suction.

In the first pair of legs the first, second, and penultimate segments are enlarged and thickened; the two intervening segments, merus and carpus, are short; the dactylus forms a curved finger tipped with a stout spine and capable of complete flexion on the robust propodus. In one or two of the succeeding pairs of legs the propodus may be slightly enlarged. The first three pairs of legs have the carpus, or antepenultimate segment, triangular, and their basal segments are directed strongly backward. In the last four pairs the carpus may be short, but is not triangular, and always distinctly separates the merus from the propodus; they are so articulated to the body that their basal segments are directed forward. The first three pairs of legs are articulated to the anterior part of the segment to which they belong, the next three near the middle of the corresponding segments, and the last pair near the posterior margin of the last segment.

The pleon is short, with the segments more or less consolidated, and the pleopods are of the normal number and form. The "operculum" is not formed as in the *Idoteidæ* and *Arcturidæ* of the uropods, but is nothing more than the enlarged and thickened first pair of pleopods, the greater part of it being formed of the external lamella, while the uropods have an entirely different and peculiar structure. They are biramous, and consist on each side of a more or less elongated, flattened, basal segment, so articulated as to lie alongside the telson, and bearing at the apex a terminal plate, the inner ramus, in the same plane with itself, while, on its upper side near the base, stands a more or less perpendicular, oval plate, the outer ramus. The telson is directed obliquely downward, and, with the uropods, forms a ciliated cup like or flower-like termination of the cylindrical body, whence the name *Anthura*, from the Greek ἄνθος, a flower, and οὐρά, a tail.

The structure of the mouth in this family has been investigated by Prof. J. C. Schiödte, to whose original papers in the Naturhistorisk Tidsskrift I have not had access. The paper on *Anthura* is translated and partly condensed in the Annals and Magazine of Natural History, where that author states that "next the *Cymothoidæ*, though as a type of a separate family, the genus *Anthura* must be placed."

The species of this family may be at once recognized by the peculiar cup-like termination of the body. This cup or "flower" is formed by the telson below, and the uropods at the sides and above; the outer rami of the latter organs being placed nearly vertically, and approaching each other on the median line above, where, however, the "flower" is more or less imperfect. Our three genera may be distinguished as follows: First five segments of pleon consolidated above, *Anthura* (p. 104); segments of pleon distinct, antennæ and antennulæ subequal, *Paranthura* (p. 108); segments of pleon distinct, antennulæ greatly enlarged in the male, *Ptilanthura* (p. 111).

Anthura Leach.

Anthura Leach, Ed. Encyc., vol. vii, p. "404" (Am. ed., p. 243), "1813–'14."

Antennulæ and antennæ short, subequal; thoracic segments not separated by constrictions; pleon with the five anterior segments consolidated above and resembling the last thoracic segment.

Our species of *Anthura* appears to agree in all generic characters with *A. gracilis* Leach upon which the genus was founded. In *A. polita*, however, the consolidated portion of the pleon is seen at the lower part of the sides to be composed of five consolidated segments, and bears the normal number of pairs of pleopods, while Bate and Westwood[*] say that "the four anterior segments are soldered closely together" in *A. gracilis*, and that "the pleopoda consist of, at least, four pairs of oval plates, strongly ciliated, on each side of the ventral surface of the basal segments of the tail." They had not, however, fresh specimens of the species, which is evidently closely related to ours.

The incubatory pouch of the females in the genus is confined to the third, fourth, and fifth segments, and is composed of three pairs of lamellæ, which overlap from behind forward, while the anterior margins of the first pair are united to the anterior part of the third segment.

Anthura polita Stimpson.

? *Anthura gracilis* DeKay, Zoöl. New York, Crust., p. 44, pl. ix, fig. 34, 1844 (*not* of Montagu and Leach).

Anthura polita Stimpson, Proc. Acad. Nat. Sci. Phil., vol. vii, p. 393, 1856.

Harger, Proc. U. S. Nat. Mus., 1879, vol. ii, p. 162, 1879.

Anthura brunnea Harger, This Report, part i, p. 572 (278), 1874.

Verrill, This Report, part i, p. 426 (132), 1874.

PLATE XI, FIGS. 68 and 69.

This species is distinguished among its allies on our coast by the nearly complete union of the basal segments of the pleon, which have together the appearance of an eighth thoracic segment. The cup or "flower" at the end of the body serves to distinguish it from other Isopoda.

---

[*] British Sessile-Eyed Crustacea, pp. 157 and 160.

The body is smooth, shining and flattened above and broadly keeled in the males below. The head is a little broader than long, deeply excavated on each side of the front for the bases of the antennulæ, and produced at the sides. The eyes are small and lateral but distinct, and are placed on the outer side of the anterior prolongations of the head, about on a line with the bases of the antennulæ. They are too indistinct in the figure, and the eye was even omitted on the right side by the engraver. The antennulæ (pl. XI, fig. 68 a) consist of a tapering three-jointed peduncle and a very short flagellum. The first peduncular segment is the largest, and is flattened above and on the inner side; the second segment is smaller, cylindrical, and provided with a comb of hair-like setæ along its outer side; the third is smaller and shorter than the second; the flagellum consists of a single very small segment, with indications of a rudimentary second segment at the end, where it is also tipped with setæ. The antennæ (pl. XI, fig. 68b) consist of a five-jointed peduncle, and a short flagellum much like that of the antennulæ. The basal segment of the peduncle is short; the second segment is the largest and is of peculiar shape, being excavated on the outer side to adapt it to the antennula, which lies in the groove thus formed, while the segment is bent upward and inward, and exposes a slender triangular area* with the point backward, between, and on a level with, the antennulæ; the next three segments are sub-cylindrical and diminish in size, and are followed by one or two small flagellar segments tipped with setæ.

The maxillipeds (pl. XI, fig. 69a) are thick and strong, and are composed of a basal quadrate segment, a little longer than broad, with its proximal external angle elided for the short, sub-triangular external lamella, and bearing two segments representing the palpus. Of these segments the first is but little smaller than the basal segment and is sub-quadrate, tapering a little at the sides beyond the middle. The terminal segment is straight at its articulation with the preceding, and nearly so along the inner side, then rounded in the remainder of the outline. The segments of the palpus are finely ciliated along their margins, except along the external margin of the first segment, where the ciliation nearly disappears; they are also provided with coarse setæ, a few of which occur on the maxilliped, near the outer distal angle. The inner maxilla (pl. XI, figs. 69 b and b′) is rather robust, and terminated by a strong tooth or spine, below which, on the inner side, is a row of smaller curved teeth. The mandibles are terminated by a horny tooth, below which is a serrulated lobe; the mandibular palpus is robust; the second segment much the longest and provided with stout setæ; the last segment with a comb of rather short setæ. The maxillipeds are of much firmer texture than the other parts of the mouth.

The first thoracic segment is the longest, and is closely adapted to the head behind so as to allow but little motion. The second segment is shorter but somewhat broader than the first, and is rather freely articulated with it, and still more freely with the third; it is carinated below, but its articulations are much less free than in the next genus. The third, fourth and fifth segments are each about the length of the second; the sixth and seventh are progressively shorter. The first pair of legs (pl. XI, fig. 68 c) are quite robust and have but little freedom of motion, being directed forward under the head and hardly capable of further lateral extension than is shown in the figure of the animal. The basis and ischium are large and articulated so as to form a curve, bringing the legs forward; the merus is short; the carpus is triangular and extends along the side of the thickened propodus for about half its length, projecting like a tooth at the end; the propodus is ovate, much thickened and armed with a tooth near the middle of the palmar margin, along which it is ciliated, as is also the carpus; the dactylus is short and stout and tipped with a slender, curved, chitinous claw about as long as the dactylus itself. The figure (pl. XI, fig. 68 c) represents the inner surface of the leg, the merus being much less conspicuous on the outer side. The second and third pairs of legs are nearly alike and much more slender than the first pair. One of the third pair is represented on plate XI, fig. 68 d. In both these pairs of legs the carpus is small and triangular and wedged in between the merus and propodus, which meet above; the merus is a little larger in the second than in the third pair, and in both pairs it is provided with a few setæ at the upper distal angle and along the opposite or palmar side, where the carpus is also armed with setæ; the dactylus bears a few very short setæ. The remaining pairs of legs are rather more slender than the second and third, and the merus is separated from the propodus above by the carpus, which is, however, short. These legs are somewhat hairy, like the preceding pairs.

The anterior part of the pleon (pl. XI, fig. 68 g), consisting of the first five segments consolidated, appears much like an eighth thoracic segment a little longer than the seventh; traces of the sutures between the segments can be seen at the sides. The last segment is distinctly articulated, a little elevated dorsally, where it is also somewhat hairy; at the lower part of the sides it is covered by a slightly projecting lobe of the preceding segment, which extends over the proximal part of the basal segment of the uropods. Distally the terminal segment is depressed at a steep angle, and is in the form of a plate, ovate and ciliated at and near the tip, where it is obtuse; the sides are nearly parallel, and it is surpassed by the uropods, which consist, on each side, of a large basal segment, carinated on the outer side and toothed at the articulation with the outer ramus, obliquely truncated at the end, where it bears a short, obtusely-triangular, ciliated, inner ramus, or lamella, in the same plane as the basal segment. The outer ramus, or lamella, forms nearly a right angle with the basal segment, and stands upon its superior outer margin. This ramus is elongate reniform in outline, being notched below for the tooth on the basal segment, and is ciliated along its free superior margin. The first pair of pleopods (pl. XI, fig. 68 e) are composed on each side of a short, quadrate basal segment supporting two rami, of which the outer is, like the basal segment, of firm texture, and acts as an operculum; in shape it is semi-oval, with the inner margin nearly straight, and is ciliated distally, and along the outer margin. The inner ramus is much smaller than the outer and of delicate texture, and, in the natural position, is covered and concealed by the outer ramus; it is slender, with nearly parallel sides, rounded at the tip, and not ciliated. In the males the second pair of pleopods (pl. XI, fig. 68 f) bears, near the middle of the inner margin of the inner ramus, a slender stylet, slightly surpassing the lamella to which it is attached.

The lamellae forming the incubatory pouch of the females are of considerable antero-posterior dimensions, and the posterior widely overlap the anterior ones, while the anterior border of the first lamella is united with the third thoracic segment, to which the lamella belongs.

Length 15–18^{mm}; breadth 1.8–2^{mm}. The color is brownish above, mottled with yellowish or honey color, lighter underneath.

This species was described as new by the present author in the first part of this report under the name *A. brunnea*, but there appears to be no sufficient reason for regarding it as distinct from Dr. Stimpson's *A. polita*. It is apparently closely related to *A. gracilis* Leach, although sufficiently distinct according to Bate and Westwood's[*] description and figures. Those authors, however, seem to have had but very poor and imperfect material on which to base their work. They figure and describe the telson and uropods as truncated and crenulated, and Montagu,[†] in his original description of the species, says that "the body is terminated by five large caudal appendages truncated at their ends."

Kröyer's[‡] descriptions and figures of *A. carinata* approach much more closely to the present species. His figure of the antennula considerably resembles ours, but in his description he gives as the relative lengths of the four segments composing it 11, 4, 3, 5. In our species the last or flagellar segment is much the shortest, as may be seen by the figure, plate XI, fig. 68 a. He further speaks of the telson as crenulated, while it is entire in *A. polita*, and his figure (Voy. en Scand., pl. 27, fig. 3 n') shows no tooth-like projection or angle on the basal segment of the uropods, as seen in a lateral view, and the corresponding margin of the outer or superior plate is destitute of the notch shown in the lateral view of these organs on plate XI, fig. 68 g. The inner ramus or lamella of the first pair of pleopods is also figured as much larger and more expanded distally than in our species, for which see plate XI, fig. 68 e. Unfortunately I have had no European specimens for comparison.

---

[*] Brit. Sess. Crust., vol. ii, p. 160, 1868.
[†] Trans. Linn. Soc., vol. ix, p. 103, pl. v, f. 6, 1808.
[‡] Naturhist. Tidssk., II, B. ii, p. 402, and Voy. en Scand., Crust., pl. xxvii, fig. 3 a–o, 1849.

This species was described by Dr. Stimpson from specimens taken at Norfolk, Va., and has since been collected by Professors Smith and Verrill at Great Egg Harbor!, N. J., in 1½ fathoms shells and mud; by the U. S. Fish Commission in Long Island Sound!, especially at Noank Harbor!, among eel-grass (*Zostera marina*) and mud; off Block Island! in 17 to 19½ fathoms sand, mud, and stones; at Vineyard Sound!, at low water and in sand, and in 1878 at Gloucester!, Mass., in mud and among algæ.

Specimens examined.

Number.	Locality.	Fathoms.	Bottom.	When collected.	Received from—	Number of specimens.	Dry. Alc.
2077	Great Egg Harbor, N.J.	1½	Shells and mud	Apr. —, 1871	Smith & Verrill		Alc.
2078	Noank Harbor, Conn		Eel-grass	Aug. 28, 1874	U. S. Fish Com.	2	Alc.
2079	...do		Mud and eel-grass	Aug. 29, 1874	...do	2	Alc.
2080	...do		Mud	Aug. 28, 1874	...do	2	Alc.
	Vineyard Sound	L. w.	Sand	Sept. 8, 1871	...do	2	Alc.
	Squan Estuary, Gloucester, Mass.		Mud	—, 1878	...do	2	Alc.
	Gloucester, Mass		Mud and algæ	—, 1878	...do	1	Alc.

Paranthura Bate and Westwood.

Paranthura Bate and Westwood, Brit. Sess. Crust., vol. ii, p. 163, 1866.

Pleon articulated, composed of six segments; thorax deeply constricted at each end of the second segment; antennulæ and antennæ subequal; palpus of maxillipeds three-jointed; inner maxillæ acicular.

The first character given above is the only one given by Bate and Westwood, who, however, mention that the pleon bears the normal number of pleopods; a character that would not distinguish our species from the other genera. The distinctly articulated flagellum of the antennulæ is provided with a partial whorl of bristles, which, however, forms only the most rudimentary approach toward the structure of those organs in the males of the following genus. The segmentation of the pleon is indistinct in the dorsal region, but is apparent at the sides when seen from above, and the pleon does not at all resemble an additional thoracic segment as in *Anthura*. Both pairs of antennæ are provided in our species with a distinctly articulated flagellum, and are of nearly equal length.

Paranthura brachiata Harger (Stimpson).

Anthura brachiata Stimpson, Mar. Inv. G. Manan, p. 43, 1853.
Verrill, Am. Jour. Sci., III, vol. v, p. 104, 1873; ibid., vol. vii, pp. 42, 411, 502, 1874; Proc. Am. Assoc., 1873, pp. 350, 357, 1874; This Report, part i, p. 511 (217), 1874.
Whiteaves, Am. Jour. Sci., III, vol. vii, p. 213, 1874; Further Deep-sea Dredging, Gulf of St. Lawrence, p. 15. "1874."
Harger, This Report, part i, p. 573 (279), 1874.
Smith and Harger, Trans. Conn. Acad., vol. iii, p. 16, 1874.
Paranthura brachiata Harger, Proc. U. S. Nat. Mus., 1879, vol. ii, p. 162, 1879.

PLATE XI, FIG. 70.

The deep constrictions, by which the second thoracic segment is separated from the first and third, serve to distinguish this species from the allied forms on our coast, and the "flower" at the end of the pleon distinguishes it from other Isopoda.

Body moniliform, with evident segments; head narrower than, and about half as long as, the first thoracic segment, flattened and quadrate above, with a groove behind a raised anterior border, wedge-shaped below, deeply emarginate on each side of the projecting front above for the bases of the antennulæ; eyes lateral, not conspicuous, extending behind the emarginations. Antennulæ (pl. XI, fig. 70 a) with the first segment large but longer than broad, flattened above; second and third segments cylindrical; flagellum of twelve or more segments in adult specimens, with the first segment short, second twice as long and the longest segment of the flagellum, which tapers from the second segment and bears on the distal end of each segment an imperfect whorl of hairs. The antennæ (pl. XI, fig. 70 b) slightly surpass the antennulæ. They have the first segment short; the second flattened on the inner side, where it is usually in contact with its fellow of the opposite side, and excavated on the outer side above to accommodate the basal segment of the antennulæ; the third segment is short; the fourth and fifth longer and cylindrical. The flagellum consists of about twelve segments, tapers from the base, and is somewhat hairy. Both the antennæ and antennulæ are a little less developed and have one or two less segments in the females. The maxillipeds (pl. XI, fig. 70 c) are elongated, with a short, oval external lamella, and a two-jointed palpus. The large basal segment of the maxilliped projects on the inner side nearly to the end of the first segment of the palpus. The palpus has its segments of about equal length and provided with a few scattered bristles. The inner maxillæ (pl. XI, figs. 70 d and d') are evident at the tip in an under view of the head; they are elongate and acicular, and minutely and sharply retro-serrate toward the tip. The three-jointed palpus of the mandibles is also conspicuous below; all three of its segments are short, and the last, which lies ordinarily between the bases of the antennæ, is flattened, oval, and provided with the usual comb of setæ.

The thorax is somewhat flattened above, carinate anteriorly below, and has the last segment much the shortest. The first segment is wider than the head and about twice its length, and is more closely united with it than are any of the thoracic segments with each other; it is strongly carinate below, especially on its anterior part, where the carina ends in a prominent tubercle; a much more slender carina bounds the flattened dorsal portion laterally. The second segment is separated from the first by a deep constriction, and is articulated so as to allow considerable motion, especially in a vertical plane; its antero-lateral angles are prominent in the form of low, rounded tubercles, and between them are two less evident tubercles on the front margin of the segment; the dorsal surface tapers behind, and is bounded laterally by carinae; below, the segment is wedge-shaped, but not carinated; behind, it is separated from the third segment by a constriction not quite as pronounced as that in front. The third segment presents two rather more evident median tubercles in front on the dorsal surface, which is defined laterally by carinae, fading away at about the middle of the segment; below, it is wedge-shaped and carinate in the males, but membranous along the median line in the females, as are the remaining segments more widely in that sex. In the males they are hard and chitinous throughout, rounded and scarcely wedge-shaped. The fourth segment is slightly longer than any of the others, and bears, near the anterior end of its dorsal surface, an oval depression with slight elongated elevations at each side. A similar structure occurs on the fifth and sixth segments, which are of decreasing length. The seventh is much the shortest thoracic segment, not being longer on the median line than the head; it is somewhat produced laterally.

The first pair of legs (pl. XI, fig. 70 e) are not as stout as in *Anthura polita*, and are more flexible; the carpus is the shortest segment, and is triangular, broader than long; the preceding segment, or merus, shows but little in an external view, but is more evident in an inner view, as shown in the figure, and is much broader than long; the propodus is much swollen proximally on its anterior or upper side; immediately in front of the end of the carpus it bears a stout tooth; the dactylus is strong, and tipped with a curved claw. In the second and third pairs of legs the carpus is triangular, but in the posterior pairs it is more elongated so as to distinctly separate the merus from the propodus.

The pleon is short, the telson triangular, acute at the apex. Uropods with the basal segment strongly carinate externally, terminal plate acutely triangular, proximal superior plate oval, curved and attached by its side, nearly meeting its fellow of the opposite side above. First pair of pleopods (pl. XI, fig. 70 f) with the external ramus semi-oval; internal ramus less firm in texture, ligulate, ciliated distally. Second pair of pleopods in the males (pl. XI, fig. 70 g) furnished with a slender stylet articulated at about the middle of the inner, posterior, lamella, and extending beyond its end. Both the lamellae are crossed by a transverse suture just beyond their middle, at the point where the stylet is attached to the inner one.

Length 28^{mm}; breadth 2.2^{mm}; females about one-third smaller. The color is usually light yellowish brown, or sometimes somewhat darker, but not as pronounced as in the other members of the family, and nearly the same throughout.

From *P. norvegica* G. O. Sars[*] our species is distinguished by the eyes, which, though inconspicuous, are present. It lacks the tubercle de-

---
[*] Chr. Vid. Selsk. Forh., 1872, p. 88, 1873.

scribed and figured by Heller on the head of *P. arctica*,† and the flagellar segments of both pairs of antennae distinguish it from *P. costana* Bate and Westwood.‡

This species was dredged by Dr. Stimpson "on a shelly and somewhat muddy bottom in twenty fathoms off the northern point of Duck Island," Bay of Fundy. It is rare south of Cape Cod, but was taken in Vineyard Sound! by the Fish Commission in 1871; also on St. George's Bank!, in 110 fathoms, mud and sand; Gulf of Maine!, down to 115 fathoms; Bay of Fundy!, down to 80 fathoms on muddy, shelly, and sandy bottoms; and off Nova Scotia!, 59 fathoms, pebbles, sand and rocks, and at other localities as detailed below. It was dredged by Mr. Whiteaves in 200 fathoms in the Gulf of Saint Lawrence, between Anticosti and the mainland of Gaspé.

Specimens examined.

Number	Locality.	Fathoms.	Bottom.	When collected.	Received from—	Number of specimens.	Dry. Alc.
2081	Vineyard Sound			—, 1871	U.S. Fish Com		Alc.
	Gulf of Maine, east from Cape Ann 140 miles.	115	Gravel	—, 1877	do	2	Alc.
	Gulf of Maine, southeast ¼ east from Cape Ann 13 miles.	53	Mud and stones	—, 1878	do	1	Alc.
2082	Gulf of Maine, near Brown's Bank.	82	Rocks and barnacles	—, 1877	do	2	Alc.
1365	George's Bank	110	Brown sand	—, 1872	Packard and Cooke.	2	Alc.
2083	Gulf of Maine, off Portsmouth 22 to 28 miles.	80-92	Soft mud	—, 1874	U.S. Fish Com.	3	Alc.
2084	Gulf of Maine	65	Mud, sand, and gravel.	—, 1874	do	2	Alc.
2087	Casco Bay, 20 miles southeast of Cape Elizabeth.	68	Mud	Aug. 12, 1873	do	1	Alc.
2088	Gulf of Maine, 27 miles off Portland	90		Aug. 20, 1873	do	1	Alc.
	Casco Bay			—, 1873	do	10	Alc.
2086	Gulf of Maine, 17 miles southeast of Monhegan Island.	72	Brown mud	—, 1873	do	3	Alc.
2005	Eastport, Me			—, 1870	A. E. Verrill.	1	Alc.
2067	do			—, 1872	U.S. Fish Com	4	Alc.
2001	Bay of Fundy, between Head Harbor and Wolves.	60	Mud	Aug. 16, 1872	do	8	Alc.
2002	Off Head Harbor	75-80	Sand and shells	—, 1872	do	8	Alc.
2093	Bay of Fundy			—, 1872	do	1	Alc.
2094	do	77		Aug. 16, 1872	do	1	Alc.
2096	Bay of Fundy, Grand Menan, New Brunswick.			—, 1870	A. E. Verrill.	3	Alc.
2098	Southeast from Cape Sable 18 to 22 miles.	56-59	Sand, gravel, and stones.	—, 1877	U.S. Fish Com.	2	Alc.

Ptilanthura Harger.

Ptilanthura Harger, Am. Jour. Sci., III, vol. xv, p. 376, 1878.

Antennulæ with the flagellum remarkably developed in the male, multiarticulate; second and succeeding antennular segments provided

---

† Denkschrift, Acad. Wiss. Wien., B. xxxv, p. [14] 38, pl. iv, figs. 9-12, 1875.
‡ Brit. Sess. Crust., vol. ii, p. 165, 1866.

with an incomplete, very dense whorl of fine slender hairs; pleon segmented, elongated; palpus of maxillipeds one-jointed.

The most important character of this genus is doubtless found in the structure of the antennulæ in the male sex. In the females the antennulæ are small, and the flagellum consists of a few slender rapidly tapering segments. They thus bear considerable resemblance to young specimens of *Anthura polita*, and being collected with them, were at first mistaken for them. They are distinguished by the larger and more conspicuous eyes, and by the more elongated and distinctly segmented pleon. In the presence of eyes our species differs from a form described by G. O. Sars, *Paranthura tenuis*, from near Stavanger, Norway, in which the males have a well-developed, eight-jointed and densely hairy or setiferous flagellum on the antennulæ.

Ptilanthura tenuis Harger.

Ptilanthura tenuis Harger, Am. Jour. Sci., III, vol. xv, p. 377, 1878; Proc. U. S. Nat. Mus., 1879, vol. ii, p. 162, 1879.

PLATES XI and XII, FIGS. 71-74.

Males of this species are at once recognized by the greatly developed antennulæ, resembling miniature bottle-brushes; females may be distinguished from the young of the other species by the conspicuous eyes; they are much smaller than the adults of the other species.

The body is smooth, flattened above, narrow at the middle, broadest at the base of the pleon. Head broader than the first thoracic segment and nearly as long, on the median line; longer than broad, narrowing to a point in front and much less acutely behind. The eyes are prominent, black, situated within the margin of the head and visible both above and below. The antennulæ in the males (pl. XII, fig. 74 a), when reflexed, attain the third thoracic segment; the first segment is large, but not longer than the second; the third is shorter than the second and followed by a short, subtriangular segment, which must be regarded as the first segment of the flagellum, although resembling the last peduncular segment much more than it does the succeeding or second flagellar segment; this segment is small at its base, but expands rapidly above and below and on the side which is next the body in the ordinary reflexed position of the antennula, and on these sides it bears, at its distal end, a fine and dense fringe of long slender hairs, which attain, when appressed, about the fifth following segment. Similar segments, to the number, in some specimens, of eighteen or twenty follow, forming an organ resembling a minute bottle brush or plume, whence the generic name. On one side, however, of the organ, which corresponds nearly with the outer or anterior side, according as the antennula is more or less reflexed, the whorl of hairs is interrupted. In the females (pl. XI, fig. 73) the antennulæ are shorter than the antennæ, with a short flagellum consisting of a small basal segment and a minute terminal one tipped with a few setæ. The antennæ (pl. XII, fig. 74 b) are short, differing little in the sexes, hardly surpassing the peduncle of the antennulæ in the males, with a short three or four jointed flagellum bearing a few hairs near the tip. The maxillipeds (pl. XI, fig. 71b) have a quadrate basal segment, somewhat emarginate externally for the subtriangular external lamella, and bearing a single suboval terminal segment, or palpus, somewhat truncate and ciliated at the tip. The inner maxillæ (pl. XI, fig. 71c) are five-toothed, one tooth being strong and terminal and the other four lateral. The mandibles bear a single-jointed palpus.

The thoracic segments are subequal in length except the last, which is but little over half as long as the others, though broader behind than any of them. They are slightly narrower than the head and margined laterally with a somewhat raised ridge. The third, fourth, and fifth have an elongate oval depression on the median line near the anterior margin. The first pair of legs (pl. XI, fig. 72) have the segments well separated, the carpus nearly equilaterally triangular, the propodus moderately thickened, and the dactylus strong and tipped with a stout claw; the carpus and propodus are bristly on their palmar margins. The remaining pairs of legs are slender and nearly equal in size.

The pleon is about as long to the tip as the last three thoracic segments. The first five segments are consolidated along the dorsum, but distinct at the sides. Each segment rises into a low broad tubercle on each side of the median line. The last segment is about as long as the preceding five, and is elongate-ovate, and obtusely pointed behind. The basal plate of the uropods is about half as long as the telson; the terminal or inner lamella is triangular-ovate, and about equals the telson. The proximal or superior lamella is narrowly semi-ovate, with an emargination on the upper side near the tip. The first pair of pleopods (pl. XI, fig. 71d) are shorter than the abdomen, and have the outer plate semi-obovate and the inner shorter, with nearly parallel sides. The second pair of pleopods (pl. XI, fig. 71e) bear, in the males, a slender straight stylet, articulated below the middle of the inner lamella and slightly surpassing it. The outer lamella is imperfectly articulated near the middle.

Length 11^{mm}; breadth 0.9^{mm}; females about one-third smaller; color brownish and more or less mottled above, lighter beneath, margined with translucent at the sides, extending on the sides of the head as far as the eyes.

This species is rare on the coast. It has been taken by the United States Fish Commission, on muddy bottom, in Noank Harbor, Long Island Sound!; off Watch Hill!, R. I., in 18 fathoms, sand; and off Block Island!, in 17 to 19½ fathoms, sand, mud, and stones; at Waquoit, Vineyard Sound!, in sand, at low water, September 8, 1871; in Casco Bay!, sand and mud, from 9 fathoms, in 1873, and by Prof. A. E. Verrill, at Grand Menan, in the Bay of Fundy! in 1870.

It is nearly related to and doubtless congeneric with *Paranthura* *tenuis* G. O. Sars,* but is at once distinguished by the presence of eyes, from which character, as distinctive, the name *P. oculata* might be applied to our species if a new trivial name should be thought necessary.

Specimens examined.

Number.	Locality.	Fathoms.	Bottom.	When collected.	Received from—	Specimens		Dry. Alc.
						No.	Sex.	
2099	Noank Harbor, Conn		Mud	—, 1874	U. S. Fish Com	1	♂	Alc.
2104	..do			—, 1874	..do	1	♀	Alc.
2100	Off Watch Hill, R. I	18	Sand	July 31, 1874	..do	1	♀	Alc.
2105	Off Block Island	17-19½	Sand, mud, and stones.	—, 1874	..do	1	♀	Alc.
2103	Vineyard Sound, Mass.	L. w.	Sand	Sept. 8, 1871	..do	1	♀	Alc.
2101	Casco Bay, Me		Mud	July 16, 1873	..do	1	♂	Alc.
2102	..do	9	Sand and mud	Aug. 4, 1873	..do	1	♂	Alc.
2106	Bay of Fundy, Grand Menan.			—, 1870	A. E. Verrill	1	♀	Alc.

XIII.—GNATHIIDÆ.

Thorax with only five pairs of legs of the normal form in the adults, and apparently consisting of only five segments; antennulæ and antennæ short, with evident distinction into peduncle and flagellum; mouth organs suctorial in the larval state, more or less aborted in the adult; pleon with its segments distinct, bearing the normal number of pleopods; uropods inserted at the sides of the base of the last segment, biramous and resembling the pleopods but of firmer texture.

This family is represented on our coast by a number of forms, all of which, however, appear to be referable to a single species, in which, contrary to what is ordinarily observed in the order, a considerable transformation occurs, especially in the males, after the young leave the incubatory pouch, and before they reach the adult form. The sexes are very unlike at maturity, but in both the thorax may be seen, by a little inspection, to consist in reality of seven segments, of which the first is united with the head, but separated from it by a sutural line near its posterior margin, while the seventh is small and resembles the segments of the pleon, which appears as if consisting of seven segments. The last thoracic segment does not bear a pair of legs. The head is large in the adult male and armed with a powerful pair of curved jaws projecting strongly forward and curved upward. The antennulæ are short and widely separated at base. The antennæ are inserted nearly below them.

The five pairs of pediform legs are ambulatory and nearly alike throughout; the propodal segments are somewhat elongate, and the dactyli weak. All the thoracic segments except the first are distinct in the male, and all are distinct in the larval forms, but the fourth and fifth

---

*Chr. Vid. Selsk. Forh., 1872, p. 89, foot-note, 1873.

(third and fourth free segments) are indistinctly separated in the adult females.

The pleon is much alike in both sexes and the young, and consists of six distinct segments, each of which bears a pair of appendages. The first five pairs of these appendages, or pleopods, are carried beneath the pleon and subserve the purposes of respiration, while they are also used in swimming. They consist of a short basal segment supporting two rami, ciliated at the tip in the young. The uropods are directed backward and are of firmer texture than the pleopods. They are ciliated near the tip.

Only a single species has yet been recognized within our limits, and the male, female, and young will be described under the specific name.

The striking sexual differences in this family have caused much confusion, the males having been referred to one genus (*Anceus*), and the females to another (*Praniza*), and even these genera have been referred to different tribes or subfamilies. The true relationship of these forms, long ago suspected by Leach, was first made known by M. Hesse,* who, however, seems not to have stated it very clearly and perhaps did not correctly apprehend it at first. His descriptions, however, of the females of *Anceus* apply to what had previously been regarded as the female of *Praniza*, although he says in the same paper that *Praniza* is only the larval state of *Anceus*, which is true only of the young, or larval forms, or the then supposed males of *Praniza*. This family has been further investigated by Bate, Westwood, and Dohrn, to whose writings the reader is referred. It may be here remarked that Bate and Westwood in their account of the structure of *Anceus*, in the second volume of the British Sessile-Eyed Crustacea, appear to have overlooked the last thoracic segment, and suppose that either the first or second segment must be wanting. Dohrn calls attention to the rudimentary (or embryonic) condition of the seventh thoracic segment as the one missing to complete the normal number, but describes and figures† as "untere" and "obere Mundextremität" ("verwandeltes erstes" and "zweites Gnathopoden Paar") what I regard as the maxillipeds and first pair of thoracic legs, or, according to Spence Bate's terminology, which Dohrn seems to have misapprehended, the maxillipeds and the first pair of gnathopods. The second pair of gnathopods are pediform as usual in the Isopoda, and are the first of the five pairs of legs. Of the five pairs of pereiopods normally present, only four are developed in the Gnathiidae. The family is thus remarkable in the order both for the transformations undergone in its development, and for the retention after all of an embryonic feature.

Having discarded the names *Anceus* and *Praniza* for reasons given below, I have also rejected the family name *Anceidæ* and substituted for it a name, suggested by Bate and Westwood and derived from that

*Ann. Sci. nat., IV, tom. ix, p. 106, 1858.
†Zeit. Wiss. Zool., xx, taf. vii, figures 24 and 25.

of the typical genus. The name *Anceidæ* should perhaps be restored in case Risso's species should not prove to be congeneric with *Gnathia termitoides* Leach, *Cancer maxillaris* Montagu.*

Gnathia Leach.

Gnathia Leach, Ed. Encyc., vol. vii, p. "402" (Am. ed., p. 240), "1813-14."
Praniza Leach, MSS.
Anceus Risso, Crust. de Nice, p. 51, 1816.

Head very large and quadrate in the male, smaller and subtriangular in the female; first pair of legs operculiform in the male, subpediform in the female; pleon much narrower than the thoracic segments, with nearly parallel sides, and a sharply triangular telson.

The name *Anceus* Risso, which has been used by modern writers for this genus, ought, according to all rules of priority, to give way to *Gnathia* Leach, as acknowledged by Bate and Westwood,[†] who, however, hesitated to restore the name on account of Kirby's coleopterous genus *Gnathium*. While the undoubted priority of the name is a sufficient reason for its re-establishment, it may be worth while to add that *Gnathia* was not restricted by Dr. Leach to either sex alone, as that author had the sagacity to "suspect that *Oniscus coeruleatus* Montagu [*Praniza coeruleata* Desm.] was the female" of *Gnathia*, and, as far as I am aware, did not publish a generic name for the Praniza-form, although the name *Praniza* was used by him as a manuscript name, and as such appears to have been published by Latreille in the Encyclopédie Méthodique, which I have not been able to consult.

Gnathia cerina Harger (Stimpson).

Praniza cerina Stimpson. Mar. Inv. G. Manan, p. 42, pl. iii, fig. 31, 1853.
Packard, Mem. Bost. Soc. Nat. Hist., vol. i, p. 296, 1867.
Verrill, Am. Jour Sci., III, vol. vi, p. 439, 1873; vol. vii, pp. 38, 41, **411**, 502, 1874; Proc. Am. Assoc., 1873, pp. 350, 354, 358, 362, 1874.
Anceus americanus, Stimpson, Mar. Inv. G. Manan, p. 42, 1853.
Gnathia cerina Harger, Proc. U. S. Nat. Mus., 1879, vol. ii, p. 162, 1879.

PLATE XII, FIGS. 75–79.

It will be convenient first to describe the male of this species and then the female and larval forms. The powerful and prominent jaws in front of the large quadrate head of the males of this small Isopod serve to distinguish it from any other on our coast.

The shape of the body is well described by Dr. Stimpson, as "regularly rectangular, abruptly narrowed at the commencement of the abdomen, which has the appearance of another very small rectangle set into the first, and of only one-third its width." It is somewhat bristly hairy, and much tuberculated and roughened above, especially on the lateral portions of the head and on the anterior thoracic segments. The head is broader than long, depressed medially in front and produced into a rounded lobe between the projecting upturned jaws. The eyes are small

---

* Trans. Linn. Soc., vol. vii, p. 65, pl. vi, fig. 2, 1804.
† Brit. Sess. Crust., vol. ii, p. 169.

and placed well forward at the sides of the head. The antennulæ (pl XII, fig. 76 a) are shorter than the head and slender, sparingly hairy, with a short, few-jointed flagellum. The antennæ (pl. XII, fig. 76 b) are also slender, with the first segment apparently composed of two united; the second segment short; the third and fourth longer, nearly cylindrical and followed by a slender few-jointed flagellum. The jaws (pl. XII, fig. 76 c) are elongate and turned upward at the apex, irregularly and bluntly toothed near the base within, and somewhat carinate on the outer side near the middle, the carina ending rather suddenly in a tooth-like process of the jaw as seen from above. The under surface of the head is deeply and broadly grooved longitudinally, and this groove is covered by what appear to be the transformed first pair of thoracic legs (pl. XII, fig. 76 d). They are in the form of a semi-oval plate on each side, attached near the base of the external side and strongly convex and ciliated on the inner side, where they overlap. This plate is truncated at the apex, where it bears a small oval lamella; on the surface of the large plate are three large, oval, semi-transparent areas. Within these plates is another pair of organs, consisting of a large basal segment and an articulated series of four flattened ciliated segments. These may be regarded as the maxillipeds, with a four-jointed palpus.

The first thoracic segment is indicated above only by a faint sutural line near the posterior margin of the large head. It is followed by five very distinct segments, of which the first two are perhaps most distinct, short, and strongly tuberculated, especially along their posterior margins. The third free segment is broader than the second, square at the sides, with two broad lateral elevations. The fourth free segment is somewhat rounded in front, with its chitinous integument apparently not calcified along the median line. The fifth free segment is narrower than the preceding and produced at the sides around the small last thoracic segment and the base of the pleon. The legs are nearly alike throughout, somewhat hairy and spiny.

The pleon is slightly dilated at the middle, with the angles of the segments salient. The last segment is acutely triangular, ciliate behind, surpassed by the uropods, which are also ciliated with a few bristles; both rami are slender, the inner a little broader than the outer. The pleopods (pl. XII, fig. 78 e) consist of two slender elongate lamellæ, the inner longer than the outer, attached to a basal segment and not ciliated in the adults of our species.

Length 4.4^{mm}; breadth 1.3^{mm}; color dirty yellowish brown above, lighter below. This form is *Anceus americanus* Stimpson.

The adult female (pl. XII, fig. 77) differs from the male principally in the following characters: The body is smooth and tapers behind and before, but is much swollen medially, where the segmentation becomes obscure, and the thoracic region seems converted into a sack for the reception of the eggs, plainly to be seen through the transparent integument. The head is comparatively small and subtriangular, emarginate in front. The eyes are placed farther back, and the large conspicuous jaws are wanting. Under the head, the first pair of legs (pl. XII, fig. 78 a) are slender, three-jointed with a minute terminal segment, and lie upon a delicate membranous plate on each side; within these are a pair of organs resembling what I have regarded as the maxillipeds of the male.

The first two free thoracic segments are short and curved somewhat around the head; the next two segments are much enlarged and nearly coalescent, and the fifth free segment is nearly similar in form to that of the males. The last thoracic segment is short and small and, as in the male, resembles a segment of the pleon.

The pleon (pl. XII, fig. 78 c) differs little from that of the male, but the angles of the segments are less salient.

Length 3–4^{mm}; breadth 1.5^{mm}. Color "pale yellowish or waxen." Dr. Stimpson was "inclined to consider" this form as the female of *Praniza cerina*.

The larval forms bear a much greater resemblance to the female than to the male but are more slender than either, the thorax being, in the smaller specimens, but little broader than the pleon. The head is broad, with large prominent eyes, and is distinct from the first thoracic segment, its posterior margin being truncated. The antennulæ have a short basal segment to the flagellum, which is followed by an elongate cylindrical segment forming about half the length of the flagellum, but bearing at its end a few short segments. The mouth organs project beyond the head, giving it an acute outline, and are evidently formed for piercing and suction. The large jaws of the adult males are, of course, wanting. The maxillipeds are slender and elongated.

The first pair of thoracic legs (pl. XII, fig. 78 b) are elongate, with the normal number of segments, a triangular carpus, and a strong curved dactylus, reminding one of the legs of the *Cymothoidæ*. The first thoracic segment is small and short and well separated from the following segments. The next two segments are quite distinct in all the forms, but usually the fourth, fifth, and sixth segments are united much as in the adult female. These forms appear to be the young females, and were described by Dr. Stimpson under the name of *Praniza cerina*; more rarely, however, specimens are found in which all the thoracic segments are distinct and somewhat resemble those of the adult male, but with their peculiarities less marked (pl. XII, fig. 79).

The pleon resembles that of the adults, but is not suddenly much narrower than the thorax. The pleopods as well as the uropods are ciliated at the tip (pl. XII, fig. 78 d).

Both these forms of young were taken from the body of a sculpin in the Bay of Fundy in 1872, and, when fresh, their bodies were bright red. In alcohol they fade to a waxy yellow.

Adult males of this species greatly resemble *Anceus elongatus* Kröyer, but his *Praniza Reinhardi* differs in its proportions of the antennary segments from *G. cerina*.

This species was described by Dr. Stimpson from females "dredged on gravelly and coralline bottoms in 20–30 fathoms in the Hake Bay," and males "dredged on a sandy bottom in 10 fathoms off Cheney's Head," Grand Menan, in the Bay of Fundy. It has been collected by the U. S. Fish Commission in Massachusetts Bay!, off Salem, 22–50 fathoms, gravel and soft mud; Gulf of Maine!, at several localities; Casco Bay!, 50 fathoms; Bay of Fundy!, in many localities, 10 to 60 fathoms, rocks, stones, and mud, and young specimens have been taken adhering to codfish and the sculpin. It was dredged by Mr. J. F. Whiteaves in the Gulf of St. Lawrence!, in 220 fathoms, mud. Further details in regard to localities are given in the subjoined table.

Specimens examined.

Number.	Locality.	Fathoms.	Bottom.	When collected.	Received from—	Specimens. No.	Sex.	Dry. Alc.
	Massachusetts Bay, 3 miles S. E. Nahant.		Mud	Aug. 31, 1879	J. H. Emerton.	3	♀	Alc.
2108	Massachusetts Bay, off Salem E. S. E. 9 to 11 miles.	22	Gravel, stones	— —, 1877	U. S. Fish Com.	3	......	Alc.
2109	Massachusetts Bay, off Salem E. S. E. 8 to 9 miles.	33	Mud	— —, 1877	....do	1	♀	Alc.
2121	Massachusetts Bay, off Salem E. S. E. 6 to 7 miles.	25–26	Gravel, stones	— —, 1877	....do	1	♀	Alc.
2110	Massachusetts Bay, off Salem E. S. E. 11 to 13 miles.	45–50	Mud	— —, 1877	....do	12	♂♀	Alc.
	Gulf of Maine, S. E. ½ S. from Cape Ann, 6 to 7 miles.	54–60	Sand, mud	— —, 1878	....do	12	♂♀ y.	Alc.
2107	Gulf of Maine between Cape Ann and Isle of Shoals.	27–36	...do	— —, 1874	....do	1	♀	Alc.
2111	Casco Bay	50		Aug. 6, 1873	....do	2	♀	Alc.
2112	......do			— —, 1873	....do	1	♀	Alc.
2113				— —, 1873	....do	10	♂♀	Alc.
2115	Eastport, Me	10–20	Rocky	— —, 1872	....do	3	♀	Alc.
2117	......do			— —, 1872	....do	3	♀	Alc.
	......do			— —, 1868	A. E. Verrill ..	3	......	Alc.
2114	Bay of Fundy		On sculpin, &c.	— —, 1872	U. S. Fish Com	12	y.	Alc.
2116	......do	25–30		— —, 1872	....do	5	♀	Alc.
2118	......do			— —, 1872	....do	5	♂♀	Alc.
2119	Bay of Fundy, off Head Harbor.	40		— —, 1872	....do	6	♂♀	Alc.
2122	Bay of Fundy	60		1870–'72	A. E. Verrill ..	00	♂	Alc.
	Off Sable Island	160	On *Lophohelia*.	— —, 1878	U. S. Fish Com.	4	♂	Alc.
2120	Gulf of Saint Lawrence	220	Mud		J. F. Whiteaves	1	y.	Alc.

XIV.—TANAIDÆ.

Respiration cephalothoracic, taking place in a cavity beneath the walls of the united head and first thoracic segment; eyes, when present, articulated; antennular flagellum single; first pair of legs enlarged and more or less perfectly chelate; pleopods natatory, ciliated, not branchial; uropods, terete, terminal, with at least one jointed ramus.

This family differs widely from all the other Isopoda, and indeed from all the sessile-eyed Crustacea, in the structure of the respiratory organs, and in the fact that the eyes, when present, are articulated with the head, or stalked, though without any proper pedicel.

I have seen species of only two genera, *Leptochelia* Dana and *Tanais* Audouin and Edwards, from within our limits. These genera are, by some authors, united under the name *Tanais*, but there seems to be ample reasons for separating them. While they agree in many characters, they differ widely from *Apseudes* Leach, which should probably be regarded as belonging to a different family not represented on our coast, and is accordingly not included in the above diagnosis.

Our representatives of the *Tanaidæ* may be further characterized as follows: The body is subcylindrical and elongated, from four or five to at least eight times as long as broad. The head and first thoracic segment are covered by the large cephalothoracic shield, which tapers somewhat in front, and is dilated behind. Its postero-lateral regions are occupied on each side by the branchial cavity, opening behind by a vertical slit, and in front by a nearly horizontal orifice. During life a lash-like organ can be seen through the body wall, in constant vibration, propelling a stream of water from behind forward through the cavity. The eyes, when present, are distinctly articulated with the head, and in the males are generally larger and more coarsely granulated than in the females. They are absent in one of our species, as in the one mentioned by Willemoes-Suhm from 1,400 fathoms in the Atlantic Ocean, off the North American coast, obtained by the Challenger expedition. They are described as indistinct in other foreign species. The antennulæ are inserted close together immediately below the vertex of the head and between the eyes. They are robust at base, and in the males may be elongated, but in the females are short, with only three or four segments and a minute rudiment of a flagellum. In neither sex have they any trace of the secondary flagellum seen in *Apseudes*. The antennæ are more slender than the antennulæ, and inserted almost directly beneath them. They are five-jointed, with the first and second segments short, the third larger and longer, the fourth and fifth slender and cylindrical, and, like the antennulæ, with indications of a flagellum. The antennæ, like the antennulæ, are tipped with bristles and bear a few scattered similar bristles on their segments.

The mouth organs are aborted in the males, at least in the genus *Leptochelia*, but in the females the mouth is protected below by a well-developed pair of maxillipeds, of which the basal segments meet at an angle forming a keel on the under surface of the head. The palpi of the maxillipeds are four-jointed, and armed with strong cilia; the last segment is strongly flexed on the penultimate. The inner maxillæ are spiny, and have the outer lobe reflexed and bearing elongated cilia at the tip. The mandibles are strong, destitute of palpi, and armed with one or two dentigerous lamellæ at the apex and a strong molar process.

The first pair of legs are robust, and in the males may be large and much elongated; they are in both sexes of our species powerful organs of prehension, being strongly chelate. Like the remaining pairs of legs, they have only five movable segments, unless an articulated spine at the extremity of the fifth segment is to be regarded as the true dactylus. On the other hand, the basal segment in many specimens presents indications of a short segment at its distal end, as if really consisting of the united basis and ischium. If this latter supposition be the true one, the hand of the first pair of legs is formed, as might be expected, of the propodus and the dactylus; the propodus is thickened and provided with a digital process stronger than the curved dactylus, which closes against it; the digital process bears toward the tip a few stout, bristly setæ. These legs are attached to the under side of the united head and first thoracic segment below the branchial cavity, and are directed forward. They are capable of but little lateral motion, and are nearly in contact below, especially toward their bases, which cover and partly conceal the organs of the mouth and the bases of the antennæ. The second pair of legs are very slender in comparison with the first, and are more slender than those that follow. Their basal segments are flattened, somewhat elongated, and usually bent with the convexity outward, in adaptation to the basal segments of the first pair of legs, which they partly embrace. The last three pairs of legs have their basal segments swollen.

The pleon consists, in our species, of five or six segments, and bears three or five pairs of strongly ciliated pleopods of the ordinary form, and fitted for swimming, and also a pair of uropods, consisting of a large basal segment bearing one or two rami. This ramus, or the inner one when there are two, is articulated and composed, in our species, of from two to six segments. The outer ramus may also consist of more than one segment. Like the antennulæ and antennæ, the uropods are provided with setæ, which are often elongate.

In the young the seventh pair of legs are not developed, and, according to Müller, the pleopods are likewise wanting and the uropods have less than the adult number of segments.

This family has been the subject of special research by Fritz Müller, Spence Bate, Dohrn, and others, to whose writings reference may be had for further description of their anatomy and development. Their proper place among the Crustacea cannot be regarded as settled, though the opinion of Fritz Müller that they represent an ancestral type of Isopoda is probably the best offered as yet. According to Dohrn, they present in their development affinities with *Asellus*, *Ligia*, and *Cuma*. Gegenbaur associates his *Tanaida* with the *Podophthalma* rather than the *Edriophthalma*.

Our species of this family are sharply divided into two genera, for which I have, after some hesitation, adoped the names *Tanais* Aud. and Edw. and *Leptochelia* Dana. I have not been able to see Audouin and Edwards' Résumé d'Entomologie, in which the genus *Tanais* is said to have been established, without description, in 1829. In the Précis d'Entomologie, by the same authors, is a figure (pl. xxix, fig. 1), apparently the same as that in the Résumé, which is there called *Tanais* de Costa. Latreille,* in 1831, characterized the genus, basing it upon *Gammarus Dulongii* Aud., figured by Savigny. Westwood,† in 1832, proposed for the same species the name *Anisocheirus*, without, however, mentioning any characters. In 1836, Templeton‡ described and figured, with evident care and accuracy, a species of this family under the name *Zeuxo Westwoodiana*. This species has, according to his figure, six segments in the pleon. Edwards, in his general work, Histoire naturelle des Crustacés, figures and describes *Tanais Carolinii* (tome iii, p. 141, pl. 31, fig. 6), and refers the figure in the Précis d'Entomologie to that species. In 1843, Rathke § described and figured *Crossurus vittatus* as a new genus and species allied to *Apseudes* and *Tanais*, but there do not seem to be any characters of importance to separate it from *T. Carolinii* Edw., and, indeed, Bate and Westwood are inclined to regard them as identical species. If, however, *T. Dulongii* be regarded as the type of the genus, there appears to be nothing but the clothing of the basal segments of the pleon to separate the two genera, and this character seems of no more than specific value, since *T. Dulongii* is described by Bate and Westwood as possessing the peculiar "branchial appendages" at the base of the fifth pair of legs. These appendages are doubtless incubatory sacs, similar to those of *T. vittatus*.

For the second genus I have hitherto used the name *Paratanais* Dana, on the ground that *Leptochelia* of the same author, although having priority, was founded upon the characteristics of the male sex. The type-species, however, of this genus, *L. minuta*, possesses all the characters of *Paratanais* that could occur in the male. *Leptochelia Edwardsii* Dana, *Tanais Edwardsii* Kröyer, moreover, belongs to the same genus, and I have adopted the name for both sexes.

The minute species, by which this family is represented on our coast, may be readily recognized by the proportionately large and strong chelate first pair of legs articulated to the united head and first thoracic segment. The two genera are distinguished by the number of segments in the pleon, which are five, with three pairs of pleopods in *Tanais* (p. 122), and six, with five pairs of pleopods in *Leptochelia* (p. 126).

Tanais Audouin and Edwards.

Tanais Audouin and Edwards, "Résumé (not Précis) d'Ent., p. 182 (without description, 1829), pl. xxix, fig. 1" (B. & W.); Précis d'Entomol., p. 46, pl. xxix, fig. 1.
Edwards, Hist. nat. des Crust., tom. iii, p. 141, 1840.
Crossurus Rathke, Fauna Norwegens, p. 35, 1843.

Antennulæ and antennæ simple; mandibles without palpi; pleon composed of five segments bearing three pairs of ciliated pleopods below,

---
* Cours d'Ent., p. 403. † Ann. Sci. nat., tome xxvii, p. 380, 1832.
‡ Trans. Ent. Soc., vol. ii, p. 203, 1836. § Fauna Norwegens, p. 35.

and a pair of simple uropods behind; eggs incubated in sacs attached near the bases of the fifth pair of legs of the females.

This genus is distinguished from the next by the structure of the pleon and the uropods as given above, and the females are, when carrying eggs or young, distinguished from all the other Isopoda by the wart-like, or sac-like, appendages of the fifth thoracic segment. Usually a small wart-like appendage is visible on each side of the inferior surface of the thorax just within the bases of the fifth pair of legs, but the size of these organs varies greatly, and in some specimens they become distended with eggs, extended lengthwise with the body and more or less coalescent, so as to form the large, bilobed incubatory pouch, as figured by Rathke. This pouch is, however, attached only to the fifth segment.

The presence of a peculiar appendage to the fifth pair of legs in this genus has been noted by various authors. Bate and Westwood figure, in the second volume of the British Sessile-Eyed Crustacea, page 122, a leg of the fifth pair with the attached pouch, which they "regard as a branchial sac similar to those existing in the Amphipoda, and consequently affording a proof of the nearer relationship of *Tanais* with that order than is possessed by any other isopodous animal." They remark further that "this appendage is wanting in some specimens, and its variable existence is probably a character of specific distinction in the group." Those authors have not, however, separated *T. vittatus* into two species on this character. Stebbing[*] mentions a specimen with eggs "as described by Rathke." Macdonald[†] figures a female with an incubatory pouch, which he briefly describes as "a membranous expansion or saccule under the thorax."

Rathke's original description is as follows: "Beide Exemplare, die ich untersuchen konnte, waren Weibchen und trugen Eier unter dem Thorax. Diese aber, die übrigens verhältnissmässig ziemlich gross waren, lagen nicht, wie bei *Idothea*, *Ligia* und vielen andern Isopoden, in einer zum Theil aus Schuppen bestehende Brüthöle eingeschlossen, sondern bildeten zwei länglichovale, dicht neben einander liegende und an der Oberfläche nur wenig eubene Massen von ziemlich beträchtlicher Grösse. Jede von ihnen war zusammengesetzt aus den Eiern und einer durchsichtigen eiweissartigen Substanz, die um jene herumgegossen war, sie wie ein Kitt zusammen hielt, und sie zugleich auch an die Bauchseite des Leibes befestigte. Es zeigten demnach jene Massen ganz dieselbe Zusammensetzung, wie die sogenannten Eiertrauben der Cyclopiden, Lernaeaden und Branchiopoden." Rathke, having had only two specimens, does not appear to have perceived the attachment of these masses at the bases of the fifth pair of legs, and of course had no opportunity to see them in various stages of development. A specimen belonging to this genus and measuring 17 millimeters in length was obtained at Ker-

---

[*] Ann. Mag. Nat. Hist., IV, xvii, p. 78, 1876.
[†] Trans. Linn. Soc., II, Zool., vol. i, p. 69, pl. xv, fig. 1, 1875.

guelen Island by Willemoes-Suhm,* who describes the sacs attached to the fifth thoracic segment and attaining, as the young develop, a diameter of three to four millimeters.

Tanais vittatus Lilljeborg (Rathke).

>   *Crossurus vittatus* Rathke, Fauna Norwegens, p. 39, pl. 1, figs. 1–7, 1843.
>   *Tanais tomentosus* Kröyer, Naturhist. Tidssk., B. iv, p. 183, 1842; ibid., II, B. ii, p. 112, 1847 : Voy. en Scand., Crust., pl. xxvii, figs. 2 a–q, "1849."
>   Lilljeborg, Ofvers. Vet.-Akad. Förh., Arg., viii, p. 25, 1851.
>   Meinert, Crust. Isop. Amph. Dec. Daniæ, p. 86, "1877."
>   *Tanais hirticaudatus* Bate, Rep. Brit. Assoc., 1860, p. 224, 1861.
>   *Tanais vittatus* Lilljeborg, Bidrag Känn. Crust. Tanaid., p. 29, 1865.
>   Bate and Westwood, Brit. Sess. Crust., vol. ii, p. 125, 1866.
>   Stebbing, Trans. Devon. Assoc., 1874, p. (7), and 1879, p. (6); Ann. Mag. Nat. Hist., IV, vol. xvii, p. 78, 1876.
>   Verrill, Am. Jour. Sci., III, vol. x, p. 38, 1875.
>   Macdonald, Trans. Linn. Soc., II, Zool., vol. i, p. 67–70, pl. xv, 1875.
>   Harger, Proc. U. S. Nat. Mus., 1879, vol. ii, p. 162, 1879.

PLATE XIII, FIGS. 81, 82.

This species is at once recognized among our Isopods by the pleon, which is beset with bristly hairs at the sides, and crossed by two rows of similar hairs near the posterior margins of its first two segments.

The body, though small, is rather robust, the length being about five times the breadth, which is greatest at the first free, in reality the second, thoracic segment. The head and united first thoracic segment is short, not longer than broad. The eyes are distinctly articulated and much less in diameter than the bases of the antennulæ. The antennulæ are shorter than the head and first thoracic segment, and are composed of three segments, of which the first is longer than the other two together, while the second and third are of about equal length; the third segment is terminated by one or two rudimentary segments, surmounted by a tuft of straight bristly setæ. Similar setæ arise from the terminal portions of the two preceding segments. The antennæ are as long as the antennulæ, but more slender, and consist of a five-jointed peduncle, somewhat setose like the antennulæ, and terminated by a rudimentary flagellum beset with setæ. The basal plates of the maxillipeds are ciliated externally, and meet each other on the median line so as to form a keel narrowing backwards; distally they become thicker and bear a four-jointed palpus, of which the second and third segments are dilated internally and ciliated, and the fourth is spatulate and ciliated at its extremity. The inner maxillæ have one of the lobes of the usual form and position, and armed with short, curved spines at the tip, while the other is bent backward and bears several elongated cilia at the tip, and by its constant motion urges a stream of water through the branchial cavity.

The first pair of legs are much enlarged and extend, in their natural position, beyond the head, and the "hand" is ordinarily directed nearly downward. The digital process of the propodus bears a broad lobe on its inner side, and an acute tooth at its extremity; at the side of the lobe

---

*Zeit. Naturges., B. xxiv, p. xvii, 1874.

is a row of setæ; the dactylus is strong, with an obtuse tooth on its inner margin. In the second pair of legs the dactylus is rather robust and tapers strongly. In the succeeding pairs of legs the dactyli become curved, and, in the posterior pairs, hooked and armed with a comb of slender teeth, while the three preceding segments are also armed with slender teeth or spines at their distal ends. The constrictions between the thoracic segments are well marked, giving the body a somewhat moniliform appearance. In breeding females, a pair of warts, or sacs of greater or less size are found attached to the under surface of the fifth thoracic segment, and containing eggs or young, according to their stage of development. These sacs often, if not usually, coalesce more or less perfectly before maturity.

The first three segments of the pleon are not narrower than the last thoracic segment, and are strongly margined, or tufted, at the sides with plumose hairs. These hairs are continued in two transverse rows, one upon the first and another on the second segment near their posterior margins, across the back of the pleon. This character is only imperfectly shown in the figure, where the transverse rows of hairs should have been more strongly indicated. The last two segments of the pleon are suddenly narrower than the first three. The last is much longer than the fourth and bears a short tooth at each side near the base. This segment may be composed of two united. The three pairs of pleopods are nearly alike (pl. XIII, fig. 82), and consist of a basal segment bearing two semi-oval lamellæ, which, as well as the basal segment, are strongly ciliated. The uropods are scarcely longer than the last two segments of the pleon, and the basal segment is comparatively small; the second segment is nearly as long as the first, the third about half as long as the second and tipped with setæ, with which the first two segments are also provided.

Length 5.5^{mm}; breadth 1.1^{mm}; color brown, mottled with lighter above; beneath, nearly white.

This species occurred on piles and among algæ and eel-grass at Noank!, Conn., in the summer of 1874, along with *Leptochelia algicola*, but in much less abundance. It was described by Rathke from Molde, on the west coast of Norway, and inhabits also the British Isles, and while the present article was going through the press I received, through the kindness of Rev. T. R. R. Stebbing, specimens from Torquay!, England, which confirm my previous determination of our species as identical with the European form. It has been found by J. D. Macdonald "in the excavated wood of piers, in company with *Limnoria* and *Chelura terebrans*." It is doubtfully identified by Bate and Westwood with a Mediterranean species, *T. Carolinii* Edw. On the authority of Lilljeborg I have regarded it as identical with *Tanais tomentosus* Kröyer, although differing in the number and proportion of the segments of the pleon, as described and figured by that author. Kröyer's specimens were from Oresund, Denmark.

Leptochelia Dana.

Leptochelia Dana, Am. Jour. Sci., II, vol. viii, p. 425, 1849; U. S. Expl. Exped., Crust., p. 800, 1853.

Paratanais Dana, Am. Jour. Sci., II, vol. xiv, p. 306, 1852; U. S. Expl. Exped., Crust., p. 798, 1853.

Antennulæ and antennæ simple; mandibles without palpi; pleon composed of six segments, bearing five pairs of ciliated pleopods below, and a pair of biramous uropods behind; incubatory pouch of the females of the normal form.

The genus *Leptochelia* was constituted by Professor Dana for a form which Fritz Müller has since shown to be the male of *Paratanais* Dana, and although so far as I know the name has not hitherto been used for any but the male forms, I see no reason why it should not be adopted instead of the later name *Paratanais*. I have therefore adopted it for the four species lately described, from our coast. Dr. Stimpson's *Tanais filum* undoubtedly belongs to the same genus, making five species within our limits, only four of which I have seen. The species that I have examined may be further characterized as follows: The body is of nearly uniform size throughout. The antennulæ are directed forward and have a large basal segment, in contact with its fellow of the opposite side at its origin, and composing about half the length of the organ in the females; but in the males this segment, though absolutely much larger than in the females, may not form more than about a third of the total length of the antennula, which is nine to twelve jointed and terminated by a well developed flagellum. The antennæ differ but little in the sexes, and are five-jointed. The organs of the mouth are abortive in the males, and the oral region is covered below by a pair of subtriangular plates, perhaps the rudiments of the maxillipeds. The second thoracic segment is shorter than those that follow it; the fifth and sixth are the longest, and the seventh is shorter than the sixth.

The pleon consists of six distinct segments, subequal in length or with the last somewhat longer than the others. These segments are smooth above, and the first five bear on their under surface each a pair of pleopods, much like those of *Tanais* (pl. XIII, fig. 82), but not ciliated on the basal segment. The last segment bears a pair of uropods, which consist of a large basal segment bearing two terete rami. Of these the outer ramus is shorter and smaller than the inner, and may consist of a single segment so small and short as to be easily overlooked; the inner ramus is larger and longer, and composed, in our species, of from two to six segments. The number of these segments appears to be of value as a specific character, but not perfectly constant.

In the females the incubatory pouch is formed, as in the order generally, by four pairs of lamellæ attached to the bases of the second, third, fourth, and fifth pairs of legs.

Leptochelia algicola Harger.

Leptochelia Edwardsii Bate and Westwood, Brit. Sess. Crust., vol. ii, p. 134, 1868
(*Tanais Edwardsii* Kröyer?).
Tanais filum Harger, This Report, part i, p. 573 [279], 1-74 (*non* Stimpson).
Verrill, This Report, part i, p. 3-4 (87), 1874.
Paratanais algicola Harger, Am. Jour. Sci., III, vol. xv, p. 377, 1878.
Leptochelia algicola Harger, Proc. U. S. Nat. Mus., 1879, vol. ii, p. 162, 1879.

PLATES XII and XIII, FIGS. 80, 83–86.

The large and strong chelate claws, six-jointed pleon, and uropods with a short, one-jointed, outer ramus and a six-jointed inner ramus, will, in general, distinguish the present species from any other Isopod on our coast.

The body is of nearly uniform size throughout, and not constricted at the articulations. The head is narrowed in front. The eyes are conspicuous and plainly articulated, and are large in the males. The antennulae in the females (pl. XIII, fig. 84 *a*) are shorter than the head and first thoracic segment, and are composed of three segments, of which the first is longer than the second and third together, and the third is slightly longer than the second, and, in some specimens, present traces of a division into two segments. The basal segment bears a short, stout seta just beyond the middle and one or two more near the tip; the second has also setae near the tip, and the third bears a tuft of half a dozen or more setae at the tip. In the males (pl. XII, fig. 80) the antennulae are about two-thirds as long as the body and usually eleven-jointed, but sometimes with one or two segments more or less than that number. The basal segment forms, in this sex, about one-third the length of the organ, and is curved from near the base so as to be convex upward; the next two segments decrease rapidly in length, and are followed usually by eight flagellar segments provided with "olfactory setae" from two to four or more to a segment. The antennae (pl. XIII, fig. 84 *b*) in both sexes are short, slender, and decurved, terminated by a tuft of setae. They appear to vary but little in the family.

The first pair of legs have the merus triangular, bringing the ischium and carpus together. In the female (pl. XIII, figs. 83 and 84 *c*) these legs, in their natural position, extend but little beyond the head; the propodus has a stout, digital process nearly in the line of its axis; this process is broadly notched near the base, then elevated into a slightly serrulate lobe, and bears at the apex a short, stout terminal tooth. Near the base of the lobe are usually two stout setae. The first pair of legs in the males are much larger and more elongated, especially in the last three segments; the carpus is elongate and cylindrical, extending about half its length beyond the head, and attaining the end of the basal antennular segment; the propodus (pl. XIII, fig. 85) is robust and has a strong, curved, and two-toothed digital process, bearing also two stout setae near the second tooth; the dactylus is also curved and provided on its inner margin with about seven short setæ springing from the bases of as many serratures; the propodus bears on its inner surface, above the origin of the dactylus, a comb, formed by a row of short setæ, and terminated at each end by a longer one. In the second pair of legs (pl. XIII, fig. 84 d) the dactylus, with its terminal spine, is not as long as the propodus, which bears two or three setæ near its tip. The third and fourth pairs of legs are shorter than the second. The last three pairs have their basal segments moderately swollen; the merus, carpus, and propodus of these legs are armed with a few spines near their distal ends; the dactyli are short.

The pleon is slightly broader near its base than the thoracic segments. The first five segments are subequal in length, the last longer and pointed behind. The uropods (pl. XIII, fig. 86) consist of a robust basal segment (b) bearing two rami, of which the outer (o) is very short and uniarticulate; the inner (i) is six-jointed, tapering from the base, with the segments of about equal length and provided with setæ near their distal ends.

Length 2.2^{mm}; breadth 0.33^{mm}; color nearly white.

It is possible that this species may prove to be identical with *L. Edwardsii* (Kröyer) Dana, although differing from Kröyer's description[*] and figures, especially in the following particulars: The peduncle of the antennula, which, according to his description and figure, consists of a short basal segment, an elongated segment, and a third short segment, has by his description the ratio to the following flagellum of five to four. The basal segment that he describes and figures was probably only the enlarged basal portion of the elongated segment, which, together with the following segment, constitutes only about three-sevenths of the length of the organ instead of five-ninths according to his description. He further describes and figures the uropod as biramous, with the inner elongated ramus composed of seven segments instead of six. Other differences could be pointed out in the proportions of the thoracic segments and the segments of the first pair of legs. Bate and Westwood[†] figure and describe a species, which they regard as *L. Edwardsii*, although their description and figures differ somewhat from Kröyer's, principally in the fact that they figure and describe the uropods as simple, saying in the generic description: "Pleopoda, five anterior pairs biramose;' posterior pair unbranched and multiarticulate;" and again under the species (p. 136), "The posterior or caudal pair of pleopoda consist of a single multiarticulate branch, of which the basal joint is larger than the terminal ones: it consists of nine or ten small articuli." They figure it on page 134 as simple, tapering from the base and seven-jointed. These authors express their indebtedness "for this interesting addition to our British fauna to the zeal and research of the Rev. A. M. Norman, who took it during the summer of

---

[*] Naturhist. Tidssk., vol. iv, p. 174, pl. ii, figs. 13–19.
[†] Brit. Sess. Crust., vol. ii, p. 134.

1865 among *Zostera* between tide marks in Belgrave Bay, Guernsey," and in the description of *Paratanais forcipatus*, on p. 139, mention in a foot-note a specimen from the same locality, "which has a pair of six-jointed anal filaments with a short one-jointed secondary filament arising from the extremity of the basal joint. Can this be the female of *Leptochelia Edwardsii* fully grown?"

Through the kindness of the Rev. Mr. Norman I have been able to examine a specimen labeled "*Leptochelia Edwardsii*, Guernsey, 1866," and do not find that it differs from our species in any characters that can be regarded as of specific value. The antennulæ have indeed only seven flagellar segments, or ten segments in all, which is also the case in some of our specimens, though eight such segments—eleven in all—is the usual number. The thoracic segments have the same proportion to each other as in our species, and the uropods agree exactly with ours in being biramous, with the outer ramus short and uniarticulate and the inner ramus six-jointed.

This is the form of uropod described and figured by Kröyer in *Tanais Savignyi*, which, as Fritz Müller has suggested, is probably the female of *T. Edwardsii* Kr. That species has, however, according to Kröyer, a five-jointed antennula, the last segment being rudimentary. I have observed among a large number of our specimens two which had the last segment divided, though scarcely longer than in the others. These specimens could hardly be distinguished from *T. Savignyi* Kröyer by any characters that I have observed. In view, however, of the great similarity of the females throughout the genus, as exemplified in the females of this species and of *L. rapax*, with both sexes of which I am familiar, I have concluded for the present to retain the specific name which I recently proposed for this species, and wait until an examination of both sexes can be had to decide the questions of specific identity.

I formerly regarded this species as identical with *Tanais filum* Stimpson, and supposed its range to extend to the Bay of Fundy. In view of the number of species now known to exist on this coast, and in the absence of any specimens from the Bay of Fundy, I now regard that as an error, and have corrected it in the American Journal of Science.

This species is rather abundant among eel-grass (*Zostera marina*) and algæ at Noank! and Wood's Holl!, and has been taken during the past summer (1879) at Provincetown!, Mass., among eel-grass, on a vessel's bottom and in old piles, in company with *Chelura terebrans* Philippi and *Limnoria lignorum* White. The specimen sent by the Rev. A. M. Norman enables me to extend its range to the Island of Guernsey!, in the British Channel.

Specimens examined.

Number	Locality	Fathoms	Bottom	When collected	Received from—	Specimens No. Sex.	Dry. Alc.
2126	Noank Harbor, Conn	.....	Eel-grass	—, 1874	U. S. Fish. Com	00 ♂ ♀	Alc.
2127	..do	.....	Eel-grass and algae.	—, 1874	..do	00 ♂ ♀	Alc.
2128	..do	.....	..do	—, 1874	..do	10 ♂	Alc.
2129	Vineyard Sound, Mass.	.....		—, 1871	..do	1 ♀	Alc.
2130	Vineyard Sound, Parker's Point	.....	L. w. on algae	—, 1875	..do		Alc.
2131	..do	.....	On piles	—, 1875	..do		Alc.
2132	..do	.....	Surface	—, 1875	..do	3 ♀	Alc.
2133	..do	.....		—, 1875	..do	8	Alc.
	Provincetown, Mass.	Lw.		Aug. 22, 1879	..do	12 ♂ ♀	Alc.
	..do	.....	Eel-grass	Aug. 23, 1879	..do	24 ♂ ♀	Alc.
	..do	1	..do	Aug. 23, 1879	..do	2 ♀	Alc.
	..do	.....	In old piles	Aug. 23, 1879	..do	1 ♂	Alc.
	..do	.....	Vessel bottom	Sept. 3, 1879	..do	2 ♂ ♀	Alc.

Leptochelia limicola Harger.

Paratanais limicola Harger, Am. Jour. Sci., III, vol. xv, p. 378, 1878.
Leptochelia limicola Harger, Proc. U. S. Nat. Mus., 1879, vol. ii, p. 163, 1879.

PLATE XIII, FIGS. 87, 88.

I have seen only females of this species, and these in general much resemble the same sex in *L. algicola* described above, but differ as follows: The eyes are small and inconspicuous, being less than half the transverse diameter of the basal antennular segment. The second segment of the antennulæ (pl. XIII, fig. 88 a) is short, only about half as long as the third. In the second pair of legs the dactylus with its terminal claw or spine is longer than the propodus, and the claw is slender and attenuated. The pleon is not wider than the segments of the thorax, and the uropods have the outer ramus two-jointed and surpassing the basal segment of the inner ramus, which is five-jointed, with the first segment long and imperfectly divided.

Length 2.5 mm. Color white in alcohol.

The specimens of this species were dredged in 48 fathoms, soft mud, in Massachusetts Bay!, off Salem, by the United States Fish Commission, in the summer of 1877.

Leptochelia rapax Harger.

Leptochelia rapax Harger, Proc. U. S. Nat. Mus., 1879, vol. ii, p. 163, 1879.

PLATE XIII, FIGS. 89, 90.

Females of this species closely resemble those of the two preceding species, but are distinguished by the following characters: The eyes are larger and more conspicuous than in *L. limicola*. The last segment of the antennulæ is scarcely longer than the preceding, instead of nearly twice as long. In the second pair of legs the dactylus is somewhat shorter, and the terminal spine less attenuated. The external ramus of the uropods consists of a single very short and small segment, shorter than the basal segment of the inner ramus, which is not elongated. The inner ramus is five-jointed instead of six-jointed, as in *L. algicola*, from which species the males are easily distinguished by the elongate and slender antennulæ and chelate legs, and by other characters, as may be seen from the following description and the figures.

The males (pl. XIII, fig. 89) are remarkable for the long, slender hand terminating the first pair of legs (pl. XIII, fig. 90). The body of the male is short and robust, and the segments are well separated by constrictions at the sides. The head with the united first thoracic segment is short and rounded, bulging strongly at the sides just behind the eyes, which are conspicuous, considerably less in diameter than the bases of the antennulæ, distinctly articulated and coarsely faceted. The antennulæ are much elongated, especially in the basal segment, which constitutes nearly half the length of the organ, and is more than one-third as long as the body; this segment is straight, swollen on the inner side near the base, then tapers gradually to the tip; the second segment is a little over one-third the length of the first and cylindrical; the third is again about one-third the length of the second, and scarcely thicker than the following flagellar segments, which vary in number from six to eight, and are usually of about equal length. In case there are eight flagellar segments the first is, sometimes at least, considerably shorter than the others. The last segment is tipped with a rudiment, and bears a few setæ. The whole number of segments, therefore, varies from nine to eleven, and if one of the flagellar segments be taken as a unit of measurement, the length of the first three segments will be approximately expressed by the numbers 9, 3.8 and 1.1. The antennæ when extended do not far surpass the middle of the basal segment of the antennulæ, and are comparatively slender; the first segment is short and somewhat expanded distally; the second is slightly longer and expanded so as to be sub-cordate; the third is short and cylindrical, equal in length to the first; the fourth is the longest segment, being longer than the first three taken together, and is slender and cylindrical, with a few setæ near the tip; the fifth is more slender and but slightly shorter than the fourth, and is tipped with a minute rudimentary terminal segment and a few setæ.

The legs of the first pair are large and much elongated. They vary somewhat in size and proportions, but are commonly, when extended, longer than the body of the animal. In these legs the segments preceding the carpus are robust but comparatively short, while the carpus is about half as long as the body, and the propodus (pl. XIII, fig. 90) is even more elongated than the carpus, and is usually strongly flexed upon it. More than half the length of the propodus is made up of the slender digital process, which bears, near the base on the inner side, a low, obtuse tooth, and a larger and more prominent one near the slender incurved tip. The dactylus (pl. XIII, fig. 90) is more than half as long as the propodus, slender, curved, and pointed, and armed with scattered, weak spinules along the inner margin. The digital process of the propodus bears also a few setæ, especially near the base of the outer tooth. The forceps thus formed are in most cases large enough to close around the body of another individual, but vary in size, being in some specimens at least one-third smaller than in others. The basal antennular segment may also be somewhat shorter than above described.

Of the thoracic segments the second (first free) segment is the shortest, and is also slightly broader than the others, and broader than the head. The third, fourth, and fifth segments increase in length progressively; the sixth is as long as the fifth; the seventh shorter. In the second pair of legs, the dactylus with its terminal claw is about as long as the propodus and nearly straight, as it is also in the third and fourth pairs, but the dactyli of the last three pairs of legs are more curved, and the basal segments somewhat swollen.

The first five segments of the pleon are of about equal length. The sixth is slightly shorter, obtusely pointed in the middle, and emarginate above the bases of the uropods, which are composed of a robust basal segment, bearing a minute outer ramus composed of a single segment tipped with setæ, and a five-jointed inner ramus, also sparingly provided with setæ. Between the uropods and below, a thin spatulate plate projects beyond the extremity of the pleon.

In length the males vary from 2.6^{mm} to 3.8^{mm}, and in breadth from 0.6^{mm} to 0.85^{mm}. The females measure in length about 2.5^{mm}; in breadth, 0.5^{mm}.

About one hundred specimens of this species, three-fourths of them females, were collected by Prof. A. Hyatt and Messrs. Van Vleck and Gardiner, in three feet of water, on muddy bottom, in the summer of 1878, at Annisquam!, Mass., and are the only specimens I have seen.

Leptochelia filum Harger (Stimpson).

 Tanais filum, Stimpson, Mar. Inv. G. Manan, p. 43, 1853.
  Packard, Mem. Bost. Soc. Nat. Hist., vol. i, p. 296, 1867.
  Harger, Am. Jour. Sci., III, vol. xv, p. 378, 1878.
 Leptochelia filum Harger, Proc. U. S. Nat. Mus., 1879, vol. ii, p. 164, 1879.

"Very minute, slender, rounded on the back, white, looking very much like a short piece of thread. Head small, and rather narrowed in front; first thoracic segment of great length; the second half as long as the third, which is about equal in length with the fourth, fifth, and sixth; the seventh being a little shorter than the sixth. The segments of the abdomen are well defined, the first five equaling each other in length, and the terminal one longer than the fifth, but narrower, and rounded behind. Antennæ short and thick, without flagellæ, with blunt tips crowned with few hairs, as are also their articulations. The inner ones are directed forward, and much the stoutest, especially toward their bases; while the outer ones are more slender and curve outward and backward. First pair of legs exceedingly thickened, with very large ovate hands and strong curved fingers. They are generally closely applied against the breast. The remaining thoracic feet are very slender, terminating in sharp, slender fingers, which in the second pair are very long and nearly straight, and in the other pairs short. The legs of the posterior pair are a little the longest and thickest. The ambulatory feet, in five pairs, are of great length and resemble those of Amphipods. The caudal stylets are in length about four-fifths that of the abdomen, and consist of four or five articles, with few hairs, each article becoming narrower, the last one with a tuft of few hairs at its extremity. Length .15 inch; breadth .02. Dredged among *Ascidiæ callosæ*, in 20 fathoms, in the Hake Bay."

I have seen no specimens corresponding fully with the above description, which is copied from Dr. Stimpson; neither have I seen any specimens of this family from the Bay of Fundy. I formerly regarded the species from Vineyard Sound as *Tanais filum* Stimpson, and that name is used in this Report, part i, p. 573 (279), where also "Bay of Fundy to Vineyard Sound" is given as its range. This error was corrected by the writer in the American Journal of Science in 1878. In the absence of specimens from the Bay of Fundy I am unable to say positively that this species is not the same as my *P. limicola*, although the number of segments in the uropods does not correspond with those of that species, and the outer ramus of the uropods, which is rather conspicuous in that species, is not mentioned at all by Dr. Stimpson. Further investigation is needed to settle this question, but the number of species known to me from the coast seems sufficient warrant for regarding this, for the present at least, as a distinct species.

Dr. Packard states that he has dredged *Tanais filum* Stimpson in the Gulf of St. Lawrence, "at Caribou Island, in eight fathoms, on a sandy bottom."

Leptochelia cœca Harger.

Paratanais cœca Harger, Am. Jour. Sci., III, vol. xv, p. 378, 1878.
Leptochelia cœca Harger, Proc. U. S. Nat. Mus., 1879, vol. ii, p. 164, 1879.

PLATE XIII, FIG. 91.

This species is at once recognized among our Tanaids by the absence of eyes. The enlarged chelate claws joined to the united head and first thoracic segment, and the six-jointed pleon serve to distinguish it as belonging to the present genus.

Body slender, elongated, and rather loosely articulated; head narrow in front, not broader than the bases of the antennulæ; eyes wanting; antennulæ distinctly four-jointed (pl. XIII, fig. 91 *a*) in the type specimen, first segment forming less than half the length of the organ, second segment longer than the third, last segment about as long as the second, slender, tapering and tipped with setæ; antennæ attaining the tip of the third antennular segment. The first pair of legs (pl. XIII, fig. 91 *b*) are robust, but less so than in the preceding species; they extend forward in the natural position about to the tips of the antennæ; they have the basal segment subquadrate, the hand or propodus less robust than the carpus, with a serrated digital process; dactylus short.

The second, or first free, thoracic segment is about two-thirds as long as the third; this in turn is about equal to the fourth and to the fifth segments; while the sixth and seventh segments are progressively somewhat shorter. The second pair of legs are scarcely more slender than the following pairs, and the basal segments are not curved around the base of the first pair.

The uropods (pl. XIII, fig. 91 c) are short, and biramous; each ramus two-jointed. The outer ramus is more slender than the inner, half its length, and bears a long bristle at the tip.

Length 2.5^{mm}; color white.

The first specimen of this species was dredged along with *L. limicola* in 48 fathoms, soft mud, Massachusetts Bay!, off Salem, in the summer of 1877, and a second specimen apparently of the same species, though differing somewhat in the antennulæ, was collected on the shore at Provincetown! during the summer of 1879. Unfortunately only a single specimen was obtained in each case, but it is very distinct from the other species of our coast. It does, however, closely approach *Tanais islandicus* G. O. Sars,* but appears to differ in the first pair of legs, which Sars describes as follows: "Pedes primi paris validi, manu sat dilatata, carpo vix angustiore, digitis palmæ longitudinem æquantibus vix forcipatis." These legs are in our species distinctly chelate, and the dactylus is much shorter than the propodus (see pl. XIII, fig. 91 b). He further says: "Uropoda sat elongata, biramosa, ramis, ambobus biarticulatis, valde inæqualibus, exteriore ne 3^{mam} quidem interioris longitudinis partem assequente." In our species the outer ramus of the uropod is about one-half as long as the inner.

GEOGRAPHICAL DISTRIBUTION.

The whole number of species enumerated is forty-six, three more than were included in my recent paper on New England Isopoda in the Proceedings of the United States National Museum. Their geographical distribution, especially on our coast, is summarized in the lists below.

The following eleven species have as yet been found only south of Cape Cod:

Scyphacella arenicola.
Actoniscus ellipticus.
Cepon distortus.
Bopyrus *species*.
Erichsonia filiformis.
Erichsonia attenuata.

Cirolana concharum.
Nerocila munda.
Ægathoa loliginea.
Livoneca ovalis.
Tanais vittatus.

---

*Archiv for Mathematik og Naturvidenskab, Band ii, p. 346 [246], 1877.

The following nineteen have been found only north of Cape Cod:

Gyge Hippolytes.
Phryxus abdominalis.
Dajus mysidis.
Janira alta.
Janira spinosa.
Munna Fabricii.
Munnopsis typica.
Eurycope robusta.
Synidotea nodulosa.
Synidotea bicuspida.

Astacilla granulata.
Cirolana polita.
Æga psora.
Syscenus infelix.
Gnathia cerina.
Leptochelia limicola.
Leptochelia rapax.
Leptochelia filum.
Leptochelia cœca.

The remaining sixteen are included in the following list as found on both sides of Cape Cod, but the letter N. is used to designate such species as are common north and rare south of the Cape, and S. signifies that the species is common at the south but rare northwards.

Philoscia vittata, S.
Jaera albifrons.
Hyarachna species.*
Chiridotea cœca.
Chiridotea Tuftsii, N.
Idotea irrorata.
Idotea phosphorea, N.
Idotea robusta.

Epelys trilobus, S.
Epelys montosus, N.
Sphæroma quadridentatum, S.
Limnoria lignorum.
Anthura polita, S.
Paranthura brachiata, N.
Ptilanthura tenuis.
Leptochelia algicola, S.

The eleven species included in the following list occur also on the coast of Europe. The British species are marked B.

Gyge Hippolytes, B.
Phryxus abdominalis, B.
Jaera albifrons, B.
Munna Fabricii.
Munnopsis typica.
Idotea irrorata, B.

Astacilla granulata.
Limnoria lignorum, B.
Æga psora, B.
Tanais vittatus, B.
Leptochelia algicola, B.

The number of Isopoda included in the present paper is considerably less than are known to inhabit Great Britain, being only about two-thirds as many as are included in Bate and Westwood's work, together with such additions to that fauna as have come to my knowledge since. As has been seen, eight, or nearly one-fifth of our marine species, are identical with those of Great Britain. The number of genera is much more nearly equal. Thirty-one marine genera are enumerated in the present paper, and of these sixteen are also British. The remaining fifteen do not appear to be represented on the British coast, but their place is filled by perhaps a rather greater number of genera. Of the families, neglecting the *Oniscidæ* as not properly included in the present paper, we come to the *Bopyridæ*, which have as yet been but little studied

* The only specimen yet known is from twenty-one miles east of Cape Cod.

on this coast. Five species only are enumerated here, two of which are also British, while Bate and Westwood enumerate twelve. A closer examination of the group may very likely add considerably to the present list.

The *Asellidæ* and *Munnopsidæ*, which Bate and Westwood would unite, have seven marine species belonging to six genera in our list, and, rejecting *Limnoria*, this number corresponds well with the British list of four genera and six species; one species, *Jæra albifrons* Leach, is identical, as are three of the genera—*Jæra*, *Janira*, and *Munna*. The more typical forms of the *Munnopsidæ* have not yet, as far as I am aware, been recognized in British waters.

The *Idoteidæ* are more numerous on our coast and appear to be more diversified than in Great Britain. I have regarded our eleven species as belonging to five different genera, while Bate and Westwood include the seven British species in a single genus. The most conservative could hardly class our species in less than three genera to one English genus, and, judging mostly from the figures and descriptions, I should be inclined to reckon three, or at least two, English genera to five on our coast in this family. One genus and species, *Idotea irrorata* Edw. (Say), is identical. Of the *Arcturidæ* a single representative has only recently been discovered within our limits, while three species, of the same genus as ours, are mentioned by Bate and Westwood, and Stebbing has since added two more species.

A single species of *Sphæroma* is the only representative on our coast of a family numbering no less than five genera and thirteen species in Bate and Westwood's volume. If the last two of these species be united as sexes of the same, and *Dynamene rubra* and *viridis* be also united, as suggested by Stebbing,* there are still left eleven representatives of this family in England to one on our coast. Our species is closely related to the British *Sphæroma serratum* Leach. *Limnoria lignorum* White is the only known representative of its family on both coasts.

The *Cirolanidæ* and *Ægidæ*, which are classed together under the latter name by most authors, have only four representatives in our limits, belonging to three genera. Two of these genera are also found in Great Britain, where they contain no less than seven species, one of which, *Ega psora* Kröyer, is identical on the two coasts. *Cirolana truncata* Norman is not included in Bate and Westwood, but these authors mention three other species belonging to as many genera in this group, making five genera and ten species from Great Britain to only three genera and four species in our waters. The *Cymothoidæ* are represented in our list by three species belonging to three genera, while Bate and Westwood say of this family, "No specimen has hitherto been satisfactorily determined as having been found in _____ seas." The Rev. A. M. Norman, however, in the Annals and M_____ _____ of Natural History for December, 1868, p. 422, mentions and brie___ describes *Anilocra medi-*

---

* Jour. Linn. Soc., Zool., vol. xii, p. 142, 1874.

terranea Leach, taken from a "small fish in rock-pools at Herm in 1865." This genus has not been found on our coast.

Of the three genera and three species of *Anthuridæ* in our list two genera are also found in Great Britain, and it is possible that one species may yet prove identical. The *Gnathiidæ* are more difficult of comparison on account of the confusion that has existed in the sexes, and the larval forms. Our specimens seem to be all referable to a single species, doubtless congeneric with the British species, the number of which may, perhaps, by a liberal estimate, be placed at three.

In the *Tanaidæ*, the genera are the same as in Great Britain, and two of our species, *Tanais vittatus* Lilj. and *Leptochelia algicola* Harger, are found on both coasts. There remain a second species of *Tanais* on the British coast, and two species of *Leptochelia* (*Paratanais* of Bate and Westwood) against four species of *Leptochelia* on our coast, as the remaining representatives of this family. The genus *Apseudes* should probably be taken to represent a family not yet found on our coast.

We have, therefore, the following list of marine families, with the genera in each, that are identical on our coast and that of Great Britain. The species have been already indicated in a preceding list:

Bopyridæ: Gyge, Phryxus, Bopyrus. Two species.
Asellidæ: Jæra, Janira, Munna. One species.
Idoteidæ: Idotea. One species.
Arcturidæ: Astacilla.
Sphæromidæ: Sphæroma.
Limnoriidæ: Limnoria. One species.
Cirolanidæ: Cirolana.
Ægidæ: Æga. One species.
Cymothoidæ.
Anthuridæ: Anthura, Paranthura.
Gnathiidæ: Gnathia.
Tanaidæ: Tanais, Leptochelia. Two species.

Further details of geographical and also of bathymetrical distribution are presented in the table on pages 439 to 441, in which the first column shows the least depth in fathoms at which each species has been collected on our coast; the second the greatest depth; and the following eighteen columns are for different localities, which may be further explained as follows: The Carolinas include Charleston, S. C., Fort Macon, N. C., and Norfolk, Va.; New Jersey includes Great Egg Harbor and Atlantic City, N. J., and Fire Island Beach, on the south shore of Long Island; Long Island Sound includes Savin Rock, New Haven, Stony Creek, or Thimble Islands, Saybrook, New London, and Norwalk, Conn.; Block Island includes Watch Hill, Block Island Sound, and the deeper water off the island; Vineyard Sound includes also Buzzard's Bay, Nantucket Sound, and off Nantucket Island; Cape Cod Bay includes Provincetown and Barnstable; Massachusetts Bay includes Salem, Nahant, Gloucester, and Annisquam, Mass.; the Gulf of Maine includes all outside of the line of 50 fathoms between Cape Cod and Nova Scotia, and extending seaward to include George's Banks; Casco Bay includes Cape Elizabeth and Quahog Bay; Bay of Fundy includes Eastport Harbor and Grand Menan, while species collected at greater depths than 50 fathoms are reckoned also in the Gulf of Maine, and the same is true of those from that depth off Nova Scotia; Nova Scotia includes also Banquerean or Quereau, Eastern and Western Banks, Miquelon Island, and the Grand Banks. Species occurring on the north shore of the Gulf of St. Lawrence are credited also to Labrador. In the last column of the table the general habitat of each species is briefly indicated.

	North Sea.	British Isles.	Arctic Ocean.	Greenland.	Labrador.	Gulf of St. Lawrence.	Nova Scotia.	Bay of Fundy.	Casco Bay.	Gulf of Maine.	Massachusetts Bay.	Cape Cod Bay.	Vineyard Sound.	Block Island.	Long Island Sound.	New Jersey.	The Carolinas.	Florida.	Least depth.	Greatest depth.	ORDINARY HABITAT.
ONISCIDÆ:																					
Philoscia vittata												—	— —		— —	— —			• • •		Shore, under rubbish.
Scyphacella arenicola																			•		Shore, in sand.
Actoniscus ellipticus																			•		Shore, under rubbish.
BOPYRIDÆ:																					
Cepon distortus	+ +	+ +	+ +	+ + +												+			12 18	160 100	Parasitic on *Gelasimus*.
Gyge Hippolytes							+	—	—	— —	— —								18	100	Parasitic on *Hippolyte*, &c.
Phryxus abdominalis				+	+											+					Parasitic on *Pandalus*, *Hippolyte*, &c.
Dajus mysidis																					Parasitic on *Mysis*.
Bopyrus, species																		—			Parasitic on *Palæmonetes*.
ASELLIDÆ:																					
Jæra albifrons	+	—		—	—	—	— —	— —	— —	—	—	—	—		—				0 0	fm 300	Under seaweed, in tide-pools, &c.
Janira alta				+				—	—	—									0	200	Muddy, stony, or shelly bottoms.
Munna Fabricii						+		—	—												Eel-grass, corallines, shells, &c.
MUNNOPSIDÆ:																					
Munnopsis typica	+		+	+		+		—		—	— —	—	—		— —	—			60 230	230 230	Muddy bottoms.
Eurycope robusta																			230	106	Do.
Hyarachna, species								—	—	—	— —	—	—		— —	—			106		Do.
IDOTEIDÆ:																					
Chiridotea cœca	+	—				+		—		—	— —	—	—		— —	—		+	0 0	fm 25	In sand at low water.
Tuftsii																			0 0	20 18	Sandy bottoms.
Idotea irrorata [1]				+ +		+								—					¼ 10	190	Algæ and eel-grass.
phosphorea																					Algæ and rocks.
robusta																					Open ocean at the surface.
Synidotea nodulosa [4]																			—	—	Rocky, stony, and sandy bottoms.
bicuspida																					

28 F

	Eel-grass, shells, &c.	Eel-grass	Eel-grass and mud, rocky and muddy bottoms	On Pycnogonum, &c.	Rocks, tide-pools, &c.	Boring in submerged timber.	Swimming; free, predatory. Do.	Parasitic on the cod and halibut. Parasitic.	Parasitic on file-fish. Mouth of Lodige. Parasitic on blue-fish, &c.	Eel-grass, mud, &c.
North Sea						+		+		
British Isles						+		+		
Arctic Ocean				+				+		
Greenland								–		
Labrador								+		
Gulf of St. Lawrence		+				+		+		
Nova Scotia		–		–		–		–		
Bay of Fundy		–				–	+			
Casco Bay		– –				–				
Gulf of Maine		–						–	– –	
Massachusetts Bay		–				+	–			–
Cape Cod Bay		–			–	–				
Vineyard Sound	–	– –			–		–		– –	–
Block Island		– –					–			
Long Island Sound	– –				–	–			– –	–
New Jersey	+ – –				–				+	–
The Carolinas		–					+		– –	+
Florida					–	–				
Greatest depth	? – –	40	220	40	10		150 130			30
Least depth	0 0 0 2	7		0			130			0

ISOPODA—Continued.
Ericksonia filiformis
Eurycope ? tricuspis
   ? stricata
   ? monstrosa
ARCTURIDÆ:
Astacilla granulata
SPHÆROMIDÆ:
Sphaeroma quadridentatum
LIMNORIIDÆ:
Limnoria lignorum
CIROLANIDÆ:
Cirolana concharum
   polita
ÆGIDÆ:
Æga psora
Syscenus infelix
CYMOTHOIDÆ:
Nerocila munda
Ægathoa loliginea
Livoneca ovalis
ANTHURIDÆ:
Anthura polita

Paranthura brachiata	20	115										+	Rocky, shelly, and muddy bottoms.
Priamthura tenuis	0	19											Sand and mud.
GNATHIIDÆ:													
Gnathia cerina	10	220							−				Rocks and mud, young parasitic on fish.
TANAIDÆ:													
Tanais vittatus	0	1											Algæ and piles.
Leptochelia algicola	0	1											Algæ and eel-grass.
limicola	48	48						−	−				Muddy bottoms.
rapax[5]								−					Algæ and eel-grass.
filum	20	20						−				+	"Among Ascidia callosa."
cæca	48	48						−					Muddy bottoms.

[1] *Munnopsis typica* has been dredged only in 66 fathoms in the Bay of Fundy within our limits, but was obtained by Mr. J. F. Whiteaves from 220 fathoms in the Gulf of St. Lawrence.
[2] *Idotea irrorata* extends also to the Baltic, the Mediterranean, the Adriatic, the Black Sea, and is even said to occur in the Caspian Sea.
[3] *Idotea robusta* was described by Krøyer from 62° north latitude, between Greenland and Iceland.
[4] *Synidotea nodulosa* extends also to British Columbia.
[5] *Limnoria lignorum* is found also in the Adriatic Sea and is said to occur in the Pacific Ocean, near San Francisco, Cal.
[6] *Leptochelia rapax* was collected at Annisquam, north of Cape Ann, and therefore not strictly within the limits of Massachusetts Bay.

NOTE.—In the above table an exclamation point is used to signify the responsibility of the author for the statement that a species has been found at the locality named. The sign + denotes that a species is said by other authors to occur at the locality under which it is used. The depths are given in fathoms; *i.*, signifies low water; *s.*, shore; *sf.*, surface. See also page 137.

LIST OF AUTHORITIES.

The present list includes only such works and articles, relating wholly or in part to Crustacea, as have been quoted, or otherwise used, in the preparation of the preceding paper, and is chiefly intended to aid in consultation of the authorities quoted. A few of the titles are necessarily given at second hand, as indicated by quotation marks in the list. The references to these works occurring throughout the article are also inclosed in quotation marks, usually with an accompanying mention of the author from whom they are taken. In all other cases the references have been made directly from the works quoted. A considerable number of authorities have not been referred to, and are omitted from the list, because at present inaccessible, and, for many of the most important works that I have been able to consult, I am indebted to the liberality of Professor S. I. Smith, who has given me the free use of his library and afforded other material aid in the preparation of the article. I have also had free access to the libraries of Professors Verrill, Marsh and Dana.

In this list, as throughout the article, the number of the series of various scientific publications is indicated by Roman numerals in capitals. As far as possible references have been made to the original paging, sometimes with that of the separata added in a parenthesis, and, in the following list, a parenthesis is used to denote that the paging is, or is supposed to be, that of the separata.

Agassiz, Alexander. Letter to C. P. Patterson, Superintendent Coast Survey, on the dredging operations of the U. S. Coast Survey steamer "Blake" during parts of January and February 1878. < Bulletin of the Museum of Comparative Zoology, vol. v, pp. 1–9. Cambridge, 1878.

Andrews, A. [Limnoria terebrans attacking telegraph cable.] < Quarterly Journal of Microscopical Science, II, vol. xv, p. 332. London, 1875.

Audouin, Jean Victor, *and* **Edwards, Henri Milne.** "Résumé d'Entomologie, ou d'Histoire naturelle des animaux articulés, complété par une iconographie de 48 planches. [2 vols.] Paris, 1828–29."

Audouin, Jean Victor, *and* **Edwards, Henri Milne.** Précis d'Entomologie ou d'Histoire naturelle des animaux articulés. Première division, Histoire naturelle des annélides, crustacés, arachnides et myriapodes, complété par une iconographie. [8vo, 70 pages, 48 plates.] Paris, 1829.

Audouin, Jean Victor. Description de l'Égypte ou recueil des observations, et des recherches qui ont été faites en Égypte pendant l'expédition de l'armée Française. Explication sommaire des planches de crustacés de l'Égypte et de la Syrie. Publiées par J. C. Savigny. Histoire naturelle, tome i, pt. 4, pp. 77–98. Paris, "1830."

Bate, C. Spence. On the British Edriophthalma. < Report of the British Association for the Advancement of Science, 1855, Reports on the state of science, pp. 18–62, pl. xii–xxii. London, 1856.

Bate, C. Spence. On Praniza and Anceus and their affinity to each other. < Annals and Magazine of Natural History, III, vol. ii, pp. 165–172, pl. vi–vii. London, Sept., 1858.

Bate, C. Spence. Crustacea. [In] List of the British marine invertebrate fauna. By Robert McAndrew. < Report of the British Association for the Advancement of Science, 1860, Reports on state of science, pp. 217–236. London, 1861.

Bate, C. Spence. Carcinological gleanings, No. ii. < Annals and Magazine of Natural History, III, vol. xvii, pp. 24-31, pl. ii. London, 1866.

Bate, C. Spence, *and* **Westwood, John Obadiah.** A History of the British sessile-eyed Crustacea. [2 vols. 8vo.] London, 1861-1868.

Beneden. *See* **Van Beneden.**

Bos, Jan Ritzema. Bijdrage tot de kennis van de Crustacea hedriophthalmata van Nederland en zijne Kusten. [8vo., 100 pages, 2 plates.] Groningen, 1874.

Bosc, Louis Augustin Guillaume. Histoire naturelle des Crustacés, contenant leur description et leurs mœurs; avec figures dessinées d'après nature. [12mo., vol. ii, 296 pages, 18 plates.] Paris, An x (1802).

Buchholz, Reinhold. Zweite Deutsche Nordpolfahrt "in den Jahren 1869 und 1870, unter Führung des Kapitän Koldewey." B. ii, Part viii, Crustaceen, pp. 262-399, pl. i-xv. Leipzig, 1874.

Buchholz, Reinhold. Mittheilungen naturwiss. Vereins v. Neu-Vorpom. u Rügen, i, pp. 1-40. *See* **Münter, Julius.**

Bullar, John Follett. The generative organs of the parasitic Isopoda. < Journal of Anatomy and Physiology, vol. xi, pp. 118-123, pl. iv. London and Cambridge, 1876.

Bullar, John Follett. Hermaphroditism among the parasitic Isopoda; reply to Mr. Moseley's remarks on the generative organs of the parasitic Isopoda. < Annals and Magazine of Natural History, IV, vol. xix, pp. 254-256. London, 1877.

Catta, J. D. Note sur quelques Crustacés erratiques. < Annales des Sciences naturelles, Zoologie, VI, tome iii, pp. 1-33, pl. i-ii. Paris, 1876.

Coldstream, John. On the structure and habits of the Limnoria terebrans, a minute crustaceous animal destructive to marine wooden erections, as piers, etc. < Edinburgh New Philosophical Journal, vol. xvi, pp. 316-334, pl. vi, 1834.

Cornalia, Emilio, *and* **Panceri, Paolo.** Osservazioni zoologico ed anatomische sopra un nuovo genere di Isopodi sedentari (Gyge branchialis). < Memorie della Reale Accademia delle Scienze di Torino, II, tom. xix, pp. 85-118, pl i-ii. Turin, 1861.

Cuvier, Georges. Le Règne Animal. *See* **Edwards, Henri Milne,** *and* **Latreille, Pierre Andre.**

Czerniavski, Voldemar. Materialia ad Zoographiam Ponticam comparatam. "Transactions of the first meeting of Russian Naturalists at St. Petersburg, 1868." pp. 19-136, pl. i-viii. "1-70."

Dalyell, John Graham. The Powers of the Creator displayed in the Creation. [3 vols., 4to, 145 plates.] London, 1851-1858.

Dana, James Dwight. Conspectus Crustaceorum, &c. Conspectus of the Crustacea of the Exploring Expedition * * continued. Crustacea Isopoda. < American Journal of Science and Arts, II, vol. viii, pp. 424-428. New Haven, 1849.

Dana, James Dwight. On the classification of the Crustacea choristopoda or tetradecapoda. < American Journal of Science and Arts, II, vol. xiv, pp. 297-316. New Haven, 1852.

Dana, James Dwight. Report on the Crustacea of the United States Exploring Expedition, under the command of Charles Wilkes, U. S. N., 1838-42. Washington, Text [4to, two parts, 1618 pages], 1853. Atlas [folio, 96 plates], 1855.

Dekay, James E. Zoology of New York or the New York Fauna. Part iv, Crustacea. [4to, 70 pages, 13 plates.] Albany, 1844.

Desmarest, Ansleme Gaetan. Malacostracés. <Dictionnaire des Sciences naturelles, tome xxviii, pp. 138-425 [56 plates]. Paris, 1825.

Desmarest, Ansleme Gaetan. Considérations générales sur la classe des Crustacés. [8vo, 446 pages, 56 plates.] Paris, 1825.

Dohrn, Anton. Untersuchungen über Bau und Entwicklung der Arthropoden. 4. Entwicklung und Organisation von Praniza (Anceus) maxillaris. <Zeitschrift für wissenschaftliche Zoologie, Band xx, pp. 55-80, taf. vi-viii.—5. Zur Kenntniss des Baues von Paranthura Costana. <Tom. cit. pp. 81-93, taf. ix. Leipzig, 1870.—7. Zur Kenntniss vom Bau und der Entwicklung von Tanais. <Jenaische Zeitschrift für Medicin und Naturwissenschaft, Band v, pp. 293-306, taf. xi-xii. Leipzig, 1870.

Duvernoy, George Louis. Sur un nouveau genre de l'ordre des Crustacés Isopodes et sur l'espèce type de ce genre, le Képone type. <Annales des Sciences naturelles, Zoologie, II, tome xv, pp. 110-122, pl. iv B. Paris, 1841.

Duvernoy, George Louis, *and* **Lereboullet, Auguste.** Essai d'une monographie des organes de la respiration de l'ordre des Crustacés Isopodes. <Annales des Sciences naturelles, Zoologie, II, tome xv, pp. 177-240, pl. vi. Paris, 1841.

Ebner, Victor von. Helleria, eine neue Isopoden-Gattung aus der Familie der Oniscoiden. <Verhandlungen k. k. zoologisch-botanischen Gesellschaft, Wien, Band xviii, pp. 95-114, pl. i. Vienna, 1868.

Edwards, Alphonse Milne. Sur un Isopode gigantesque des grandes profondeurs de la mer. <Comptes Rendus, tome lxxxviii, pp. 21-23. Paris, 1879.
  Translated in the Annals and Magazine of Natural History, V, vol. iii, pp. 241-243. London, 1879.

Edwards, Henri Milne. "Résumé d'Entomologie" *and* Précis d'Entomologie. *See* **Audouin, Jean Victor.**

Edwards, Henri Milne. Annotations in Histoire naturelle des animaux sans vertèbres, par J. B. P. A. de Lamarck, 2me Edit., tome v, 8vo. Paris, 1838.

Edwards, Henri Milne. Histoire naturelle des Crustacés, comprenant l'anatomie, la physiologie et la classification de ces animaux. [8vo, 3 vols. text, 1 vol. plates.] Paris, tome i, 1834, tome ii, 1837, tome iii, 1840.
  Published as a part of the Suites à Buffon.

Edwards, Henri Milne. Le Règne Animal distribué d'après son organisation, par Georges Cuvier. Les Crustacés avec une atlas. [Crochard edition, text 4to, 278 pages, atlas with 87 plates.] Paris, "1849."

Edwards, Henri Milne. Observations sur le squelette tégumentaire des Crustacés Décapodes et sur la Morphologie de ces Animaux. <Annales des Sciences naturelles, Zoologie, III, tome xvi, pp. 224-291, pl. viii-xi. Paris, 1851.

Edwards, Henri Milne. Rapport sur un travail de M. Hesse relatif aux metamorphoses des Ancées et des Caliges. <Annales des Sciences naturelles, Zoologie, IV, tome ix, pp. 89-92. Paris, 1858.

Eichwald, Eduard von. "Fauna Caspio-Caucasie illustrationes universae. <Nouveaux Mémoires de la Société Impériale des Naturalistes de Moscou, vol. vii. Moscow, 1842."

Fabricius, Johann Christian. Entomologia Systematica emendata et aucta secundum classes, ordines, genera, species, adjectis synonimis, locis, observationibus, descriptionibus. [8vo, 4 vols., vols. i and iii in two parts]. Hafniae (Copenhagen) 1792-1794. Index alphabeticus. [175 pages]. 1796.

Fabricius, Johann Christian. Supplementum Entomologiæ Systematicæ. [8vo, 572 pages.] Hafniae (Copenhagen) 1798. Index alphabeticus. [53 pages.] 1799.

Fabricius, Otho. Fauna Groenlandica. [8vo, 450 pages, 1 plate.] Copenhagen, 1780.

Fleming, John. Crustacea. < Encyclopædia Britannica, 7th edition, vol. vii, pp. 497–504, pl. clxxx–clxxxi, 4to. Edinburgh (1842).

Fraisse, Paul. Die Gattung Cryptoniscus Fr. Müller (Liriope Rathke). < Arbeiten aus dem Zoologisch-zootomischen Institut in Würzburg, Band iv, pp. 239–296, taf. xii–xv. 1878.

Gaimard, Paul. Voyages en Scandinavie, etc. *See* **Kröyer, Henrik.**

Gegenbaur, Carl. Elements of Comparative Anatomy. Translated by F. Jeffrey Bell, the translation revised by E. Ray Lankester. [8vo, 645 pages.] London, 1878.

Geoffroy, Étienne Louis. "Histoire abrégée des Insectes qui se trouvent aux environs de Paris, dans laquelle ces animaux sont rangés suivant un ordre méthodique. Paris, 1762, 1800, 2 vols., 4to."

Goodsir, Henry D. S. On two new species of Leachia, with a plate. < Edinburgh new Philosophical Journal, vol. xxxi, pp. 309–313, pl. vi. 1841.

Gosse, Philip Henry. A Manual of Marine Zoology for the British Isles. [Two parts, small 8vo.] London, 1855–1856.

Gould, Augustus Addison. Crustacea [List of, in Massachusetts]. < Report on the Geology, Mineralogy, Botany, and Zoology of Massachusetts. 2d edition. By Edward Hitchcock. pp. 548–550. Amherst, 1835.

Gould, Augustus Addison. Report on the Invertebrata of Massachusetts, comprising the Mollusca, Crustacea, Annelida and Radiata. [8vo, 373 pages, 15 plates.] Boston, 1841.

Griffith, Edward, *and* **Pidgeon, Edward.** The Classes Annelida, Crustacea and Arachnida arranged by the Baron Cuvier, with supplementary additions to each order. [8vo, 540 pages, 59 plates.] London, 1833.

Grube, Adolph Eduard. Ein Ausflug nach Triest und dem Quarnero. Beiträge zur Kenntniss der Thierwelt dieses Gebietes. [8vo, 175 pages, 5 plates.] Berlin, 1861.

Guérin Méneville, Felix Edouard. Iconographie du Règne Animal de Cuvier. Avec un texte descriptif mis au courant de la science. Crustacés. [8vo, 48 pages, 35 plates.] Paris, 1829–1843.

Guérin Méneville, Felix Edouard. Expédition Scientifique de Morée, Section des Sciences physiques, tome iii, pt. i, Zoologie, section ii. Des Animaux articulés. Crustacés, pp. 30–50. pl. xxvii. Paris, 1832.

Harger, Oscar. The sexes of Sphæroma. < American Journal of Science and Arts, III, vol. v, p. 314. New Haven, 1873.

Harger, Oscar. On a new genus of Asellidæ. < American Journal of Science and Arts, III, vol. vii, pp. 601–602. New Haven, 1874.

Harger, Oscar. This Report, part i, pp. 569–573. *See* **Verrill, Addison Emory.**

Harger, Oscar. Trans. Conn. Acad., vol. iii, pp. 1–57, *and* This Report, part ii, pp. 657–661. *See* **Smith, Sidney Irving.**

Harger, Oscar. Description of Mancasellus brachyurus, a new fresh water Isopod < American Journal of Science and Arts, III, vol. xi, pp. 304–305. New Haven, 1876.

Harger, Oscar. Descriptions of new genera and species of Isopoda, from New England and adjacent regions. < American Journal of Science and Arts, III, vol. xv, pp. 373–379. New Haven, 1878.

Harger, Oscar. Notes on New England Isopoda. < Proceedings of the United States National Museum, 1879, vol. ii, pp. 157–165. Separata, Washington, 1879. [List of Bopyridæ by Prof. S. I. Smith.]

Heller, Camill. Carcinologische Beiträge zur Fauna des adriatischen Meeres. < Verhandlungen der k. k. zoologisch-botanischen Gesellschaft in Wien, Band xvi, pp. 723–760. Vienna, 1866.

Heller, Camill. Die Crustaceen, Pycnogoniden und Tunicaten der k. k. Osterr-Ungar Nordpol-Expedition. < Denkschriften der mathematisch-naturwissenschaftlichen Classe der kaiserlichen Academie der Wissenschaften, Band xxxv, pp. 25–46, taf. i–v. Vienna, 1875.

Hesse, Eugène. Mémoire sur les Pranizes et les Ancées (extrait). < Annales des Sciences naturelles, Zoologie, IV, tome ix, pp. 93–119. Paris, 1858.

Hesse, Eugène. Memoir on the Pranize and Ancei (abstract). < Annals and Magazine of Natural History, III, vol. xiv, pp. 405–417. London, 1864.

Hesse, Eugène. Mémoire sur les Pranizes et les Ancées. < Mémoires présentés par divers savants à l'Académie des Sciences de l'Institut Impérial de France, tome xviii, pp. 231–302, pl. i–iv. Paris, 1868.

Hesse, Eugène. Observations sur des Crustacés rares ou nouveaux des côtes de France. 15ᵐᵉ article. Description d'un nouveau crustacé appartenant au genre Limnorie. < Annales des Sciences naturelles, Zoologie, V, tome x, pp. 101–121, pl. ix. Paris, 1868.

Hitchcock, Edward. A Catalogue of the animals and plants of Massachusetts, VI, Crustacea. < Report on the Geology, Mineralogy, Botany, and Zoology of Massachusetts, pp. 553–564. Amherst, 1833.

Hope, Frederick William. "Observations on the ravages of Limnoria terebrans, with suggestions for a preventative against the same. < Transactions of the Entomological Society, vol. i, pp. 119, 120. London, 1836."

Hope, Frederick William. Catalogo dei Crostacei Italiani e di Molti Altri del Mediterraneo. [8vo, 48 pages, 1 plate.] Naples, 1851.

Huxley, Thomas Henry. A Manual of the anatomy of invertebrated animals. [8vo, 698 pages.] London, 1877. American edition [12mo, 596 pages], New York, 1878.

Johnston, George. Contributions to the British Fauna. < The Edinburgh Philosophical Journal, vol. xiii, pp. 218–222, 1825.

Johnston, George. Illustrations in British Zoology. < London's Magazine of Natural History, vol. viii, pp. 494–498. London, 1835. Ibid., vol. ix, pp. 79–83, 1836.

Jones, John Matthew. Notes on the marine zoology of Nova Scotia. < Proceedings and Transactions of the Nova Scotian Institute of Natural Science of Halifax, N. S., vol. ii, 1869–70, part iv, pp. 95–99. Halifax, 1870.

Jones, Thomas Rupert. Manual of the natural history of Greenland. *See* **Lütken, Christian Friedrich.**

Kinahan, John Robert. Analysis of certain allied genera of terrestrial Isopoda; with description of a new genus and a detailed list of the British species of Ligia, Philougria, Philoscia, Porcellio, Oniscus and Armadillidium. < Natural History Review, 1857, Proceedings of Societies, pp. 258–282, pl. xix–xxii. Dublin, 1857.

Kingsley, John Sterling. Bulletin of the Essex Institute, vol. ix, pp. 103-108. See **Streets, Thomas Hale.**

Kingsley, John Sterling. Notes on New England Isopoda, [by O. Harger, Notice of.] < American Naturalist, vol. xiv, pp. 120-121. Philadelphia, 1880.

Kirby, William, *and* **Spence, William.** "An Introduction to Entomology, or Elements of the natural history of Insects. 5th and 6th editions. London."

Kröyer, Henrik. Grönlands Amfipoder. < Kongelige Danske Videnskabenes Selskabs naturvidenskabelige og mathematiske Afhandlinger, vol. vii, pp. 229-326, (1-98) pl. i-iv. Copenhagen, 1838.

Kröyer, Henrik. Munna, en ny Kræbsdyrslægt. < Naturhistorisk Tidsskrift, Bind ii, pp. 612-616, pl. vi, figs. 1-9. Copenhagen, 1839.

Kröyer, Henrik. Nye Arter af Slægten Tanais. < Naturhistorisk Tidsskrift, Bind iv, pp. 167-168, pl. ii. Copenhagen, 1842.

Kröyer, Henrik. Monografisk Fremstilling af Slægten Hippolyte's nordiske Arter. < Kongelige Danske Videnskabenes Selskabs naturvidenskabelige og mathematiske Afhandlinger, vol. ix, pp. 211-360, pl. i-vi. Copenhagen, 1842.

Kröyer, Henrik. Karcinologiske Bidrag. < Naturhistorisk Tidsskrift, II, Bind ii, pp. 1-123, 1846, and pp. 366-446, 1847. Copenhagen, 1846-7.

Kröyer, Henrik. Voyages en Scandinavie en Laponie, au Spitzberg et aux Féröe, Zoologie, Crustacés. (Publiées sous la direction de M. Paul Gaimard). [40 folio plates.] Paris, 1849.

Lamarck, Jean P. B. A. de M. de. Histoire naturelle des animaux sans vertèbres. 2me Edit. Revue et augmentée de notes par MM. G. P. Deshayes et H. Milne Edwards.

Latreille, Pierre Andre. Histoire naturelle générale et particulière des Crustacés et des Insectes. [8 vo, 14 vols. text, 1 vol. plates.] Paris, An x-xiii (1802-1805).

Latreille, Pierre Andre. Genera Crustaceorum et Insectorum secundum ordinem naturalem in familias disposita, iconibus exemplisque plurimis explicata. [2 vols., 16 plates.] Paris, 1806-1807.

Latreille, Pierre Andre. *Entomologie ou histoire naturelle des Crustacés, des Arachnides et des Insectes. < Encyclopédie méthodique. Paris, 1789-1825."

Latreille, Pierre Andre. Le Règne Animal distribué d' après son Organisation par M. Le Baron Cuvier, tome iv, Crustacés, Arachnides et partie des Insectes. [8vo, 584 pages.] Paris, 1829.

Latreille, Pierre Andre. Cours d'Entomologie, ou de l'Histoire naturelle des Crustacés des Arachnides, des Myriapodes et des Insectes. [8vo, 568 pages, with atlas of 24 plates.] Paris, 1831.

Latrobe, Benjamin Henry. A drawing and description of the Clupea tyrannus and Onisens prægustator. < Transactions of the American Philosophical Society, vol. v, pp. 77-81, pl. i. Philadelphia, 1802.

Leach, William Elford. Crustaceology. < Edinburgh Encyclopædia, vol. vii. Edinburgh, " 1813-14."
I have seen only an American edition, in which the article is on pp. 221-277.

Leach, William Elford. A tabular view of the external characters of four classes of animals which Linné arranged under Insecta; with the distribution of the genera composing three of these classes into orders, etc., and descriptions of several new genera and species. < Transactions of the Linnean Society of London, vol. xi, pp. 306-400. London, 1815.

Leach, William Elford. Cymothoadées. <Dictionnaire des Sciences naturelles, tome xii, pp. 338-354. Paris, 1818.

Leidy, Joseph. Contributions toward a knowledge of the marine invertebrate fauna of the coasts of Rhode Island and New Jersey. <Journal of the Academy of Natural Science, II, vol. iii, pp. 135-152, pl. x-xi. Philadelphia, 1855.

Leidy, Joseph. Notices of some animals on the coast of New Jersey. <Proceedings of the Academy of Natural Sciences of Philadelphia, 1879, pp. 198-199. 1879.

Lenz, Heinrich. Die wirbellosen Thiere der Travemünder Bucht. Theil I = Anhang I zu dem Jahresberichte 1874-1875 der Kommission zur wissenschaftlichen Untersuchung der deutschen Meere in Kiel. [24 pages, 2 plates.] Berlin, 1878.

Lereboullet, Auguste. Annales des Sciences naturelles, Zoologie, II, tome xv, pp. 177-240. *See* **Duvernoy, George Louis.**

Lilljeborg, Wilhelm. Bidrag till den högnordiska hafsfaunan. <Œfversigt af Kongl. Vetenskaps-Akademiens Förhandlingar, Arg vii, pp. 82-88. Stockholm, 1850.

Lilljeborg, Wilhelm. Norges Crustacéer. <Œfversigt af Kongl. Vetenskaps-Akademiens Förhandlingar, Årg viii, pp. 19-25. Stockholm, 1851.

Lilljeborg, Wilhelm. Hafs-Crustaceer vid Kullaberg. <Œfversigt af Kongl. Vetenskaps-Akademiens Förhandlingar, Årg ix, pp. 1-13. Stockholm, 1852.

Lilljeborg, Wilhelm. Bidrag till kännedomen om de inom Sverige och Norrige förekommande Crustaceer af Isopodernas underordning och Tanaidernas familj. [8vo, 32 pages.] Upsala, 1865.

Linné, Carl von. "Fauna Suecica, sistens animalia Sueciæ Regni Quadrupedia, Aves, Amphibia, Pisces, Insecta, Vermes, distributa per classes et ordines, genera et species. Editio altera, 8vo. Stockholm, 1761."

Linné, Carl von. Systema Naturæ per Regna tria Naturæ, secundum classes, ordines, genera, species cum characteribus, differentiis, synonymis, locis. Tomus I, Ed. 12 reformata. Holmiae (Stockholm), 1766-1767.

Lockington, William Neale. Description of seventeen new species of Crustacea. <Proceedings of the California Academy of Sciences, vol. vii, pp. 41-48 (1-8). San Francisco, 1877.

Lucas, Hippolyte. Histoire naturelle des animaux articulés. Crustacés. [pp. 1-88, 8 plates.] <Exploration Scientifique de l'Algérie pendant les années, 1840-1842. Sciences physiques, Zoologie, I. Paris, 1849.

Lütken, Christian Friedrich. Nogle Bemærkninger om de nordiske Æga-arter samt om Æga-slægtens rette Begrændsning. <Videnskabelige Meddelelser fra den naturhistoriske Forening i Kjöbenhavn, Aaret 1858, pp. 65-78, pl. 1 A. 1859.

Lütken, Christian Friedrich. Om visse Cymothoagtige Krebsdyrs Ophold i Mundhulen hos forskjellige Fiske. <Videnskabelige Meddelelser fra den naturhistoriske Forening i Kjöbenhavn, Aaret 1858, pp. 172-179. Copenhagen, 1859.

Lütken, Christian Friedrich. Tillæg til „Nogle Bemærkninger om de nordiske Æga-arter samt om Æga-slægtens rette Begrændsning"—Om Æga tridens Leach og Æga rotundicauda Lilljeborg samt om slægterne Acherusia og Ægacylla. <Videnskabelige Meddelelser fra den naturhistoriske Forening i Kjöbenhavn, Aaret 1860, pp. 175-183 (1-9). Copenhagen, 1861.

Lütken, Christian Friedrich. Ibid., 1861, pp. 274-276. *See* **Steenstrup, Japetus.**

Lütken, Christian Friedrich. The Crustacea of Greenland. <Manual of the natural history, geology and physics of Greenland and the neighbouring regions; prepared for the use of the Arctic expedition of 1875, by T. Rupert Jones, pp. 146-165. London, 1875.

Macdonald, John Denis. On the external anatomy of Tanais vittatus occurring with Limnoria and Chelura terebrans in excavated pier-wood. <Transactions of the Linnean Society, II, Zoology, vol. i, pp. 67-71, pl. xv. London, 1875.

M'Intosh, William Carmichael. On the invertebrate marine fauna and fishes of St. Andrews. <Annals and Magazine of Natural History, IV, vol. xiii, pp. 140-145, 204-221, 302-315, 342-357, 420-432, vol. xiv, pp. 68-75, 144-155, 192-207, 258-274, 337-349, 412-425. London, 1874.

Marcusen, Johann. Zur Fauna des schwarzen Meeres. <Archiv für Naturgeschichte, Jahrgang xxxiii, Band i, pp. 357-363. Berlin, 1867.

Mayer, Paul. Carcinologische Mittheilungen. VI. Ueber den Hermaphroditismus bei einigen Isopoden. <Mittheilungen aus der Zoologischen Station zu Neapel, B. i, pp. 165-179, pl. v. Leipzig, 1879.

Meinert, Fr. Crustacea Isopoda, Amphipoda et Decapoda Daniæ; Fortegnelse over Danmarks Isopode, Amphipode, og Decapode Krebsdyr. "Naturhistorisk Tidsskrift, III," pp. 57-248. Copenhagen, "1877."

Metzger, Adolf. Die wirbellosen Meeresthiere der ostfriesischen Küste. Ein Beitrag zur Fauna der deutschen Nordsee. <Zwanzigster Jahresbericht der naturhistorischen Gesellschaft zu Hannover, pp. 22-36. Hannover, 1871.

Metzger, Adolf. Nordseefahrt der Pommerania—"Zoologische Ergebnisse der Nordseefahrt, X. Crustaceen aus den Ordnungen Edriophthalmata u. Podophthalmata, taf. vi. Aus Jahresbericht der Commission zu wiss. Untersuchung des deutschen Meer, im Kiel, Jahre 1872-1873." Berlin, 1875.

Miers, Edward John. List of the species of Crustacea collected by the Rev. A. E. Eaton at Spitzbergen in the summer of 1873, with their localities and notes. <Annals and Magazine of Natural History, IV, vol. xix, pp. 131-140. London, 1877.

Miers, Edward John. Report on the Crustacea collected by the naturalists of the Arctic Expedition in 1875-1876. <Annals and Magazine of Natural History, IV, vol. xx, pp. 52-66 and 96-110, pl. iii-iv. London, 1877.

Milne-Edwards. *See* **Edwards, Alphonse Milne** *and* **Henri Milne.**

Möbius, Karl. Die wirbellosen Thiere der Ostsee. <"Bericht über die Expedition zur physikalisch-chemischen und biologischen Untersuchung der Ostsee im Sommer 1871 auf S. M. Avisodampfer Pommerania." pp. 97-144. Kiel, 1873.

Möbius, Karl. On the invertebrate animals of the Baltic. <Annals and Magazine of Natural History, IV, vol. xii, pp. 81-89. London, 1873.
Translated by W. S. Dallas from the preceding.

Mohr, Nicholas. Forsøg til en Islandisk Naturhistorie, med adskillige œkonomiske samt andre Anmærkninger. [8vo, 413 pages.] Copenhagen, 1786.

Montagu, George. Description of several marine animals found on the south coast of Devonshire. <Transactions of the Linnean Society of London, vol. vii, pp. 61-85, tab. vi-vii, 1804. Ibid., vol. ix, pp. 81-114, tab. ii-viii. 1808.

Montagu, George. Descriptions of several new or rare animals, principally marine, discovered on the south coast of Devonshire. <Transactions of the Linnean Society of London, vol. xi, pp. 1-26, tab. i-v. London, 1815.

Moore, Edward. On the occurrence of Teredo navalis and Limnoria terebrans in Plymouth Harbour. <Magazine of Natural History, new series, vol. ii, pp. 206-210. London, 1838.

Moore, Edward. Limnoria terebrans in Plymouth Harbour. < Magazine of Natural History, new series, vol. iii, pp. 195–197. London, 1839.

Moore, Edward. Catalogue of the malacostracons Crustacea of South Devon. < Magazine of Natural History, new series, vol. iii, pp. 284–294. London, 1839.

Moseley, Henry Nottidge. Remarks on observations by Capt. Hutton, Director of the Otago Museum, on Peripatus novæ-zealandiæ, with notes on the structure of the species. < Annals and Magazine of Natural History, IV, vol. xix, pp. 85–91. London, 1877.

Moseley, Henry Nottidge. Hermaphroditism in the parasitic Isopoda. Further remarks on Mr. Bullar's papers on the above subject. < Annals and Magazine of Natural History IV, vol. xix, pp. 310, 311. London, 1877.

Müller, Friedrich. Bemerkungen zu Zaddach's Synopseos Crustaceorum Borussicorum prodromus. < Archiv für Naturgeschichte, Jahrgang, xiv, Band i, pp. 62–64, pl. iv. Berlin, 1848.

Müller, Friedrich [Fritz]. Ueber den Bau der Scheerenasseln (Aselliotes hétéropodes M. Edw.). < Archiv für Naturgeschichte, Jahrg. xxx, B. i, pp. 1–6. Berlin, 1864.

Müller, Friedrich [Fritz]. Facts and Arguments for Darwin. Translated by W. S. Dallas. [8vo., 144 pages.] London, 1869.

Müller, Friedrich [Fritz]. Bruchstücke für Naturgeschichte der Bopyriden. < Jenaische Zeitschrift für Medicin und Naturwissenschaft, Band vi, pp. 53–73, taf. iii–iv. Leipzig, 1871.

Münter, Julius, *and* **Buchholz, Reinhold.** "Ueber Balanus improvisus (Darw.) var. gryphicus (Münter). Beitrag zur carcinologischen Fauna Deutschlands. < Mittheilungen d. naturwissensch. Vereins von Neu-Vorpommern u. Rügen, i, pp. 1–10, 2 plates. Berlin, 1869."

Norman, Alfred Merle. Reports of deep-sea dredging on the coast of Northumberland and Durham, 1862–64. Report on the Crustacea. < Natural History Transactions, Northumberland and Durham, vol. i, pp. 12–29, "1865."

Norman, Alfred Merle. Report of the committee appointed for the purpose of exploring the coasts of the Hebrides by means of the dredge. Part ii, on the Crustacea, Echinodermata, Polyzoa, Actinozoa and Hydrozoa. < Report of the British Association for the Advancement of Science for 1866, Reports on the State of Science, pp. 193–206. London, 1867.

Norman, Alfred Merle. Preliminary report on the Crustacea, Molluscoida, Echinodermata, and Cœlenterata, procured by the Shetland dredging committee in 1867. < Report of the British Association for the Advancement of Science for 1867, Reports on the State of Science, pp. 437–441. London, 1868.

Norman, Alfred Merle. On two Isopods belonging to the genera Cirolana and Anilocra new to the British Islands. < Annals and Magazine of Natural History, IV, vol. ii, pp. 421–422, pl. xxiii. London, 1868.

Norman, Alfred Merle. Last Report on dredging among the Shetland Isles, Part ii, Crustacea, etc. < Report of the British Association for the Advancement of Science for 1868, Reports on the State of Science, pp. 247–336 and 344–345. London, 1869.

Norman, Alfred Merle. Crustacea, Tunicata, Polyzoa, Echinodermata, Actinozoa, Foraminifera, Polycistina, and Spongida in "Preliminary Report of the Biological Results of a Cruise in H. M. S. 'Valorous' to Davis Strait in 1875." By J. Gwyn Jeffreys. < Proceedings of the Royal Society, vol. xxv, pp. 202–215. London, 1876.

Œrsted, Anders Sandöe. Beretning om en Excursion til Trindelen, en Alluvial-
dannelse i Odensef jord, i Esteraaret 1841, d. 19de Octbr. < Naturhistorisk
Tidsskrift, Bind iii, pp. 552-569, tab. viii. Copenhagen, 1841.

Owen, Richard. "The Zoology of Captain Beechey's Voyage * * * to the Pacific
Ocean and Behring's Straits, performed in H. M. Ship Blossom * * * in the
years 1825-28." Crustacea, pp. 77-92, pl. xxiv-xxviii. London, "1839."

Packard, Alpheus Spring. A list of animals dredged near Caribou Island, Southern Labrador, during July and August, 1860. < Canadian Naturalist and Geolo-
gist, vol. viii, pp. 401-429, pl. i-ii. Montreal, 1863.

Packard, Alpheus Spring. Observations on the glacial phenomena of Labrador
and Maine, with a view of the recent invertebrate fauna of Labrador. < Memoirs
of the Boston Society of Natural History, vol. i, pp. 210-303, pl. vii-viii. Boston,
1867.

Packard, Alpheus Spring. On the Crustaceans and Insects [in] The Mammoth
Cave and its inhabitants, by the editors. < American Naturalist, vol. v, pp. 744-
761. Salem, 1871.

Packard, Alpheus Spring. On the cave fauna of Indiana. < Fifth Annual Report
of the Trustees of the Peabody Academy of Science, pp. 93-97. Salem, 1873.

Panceri, Paolo. Mem. Accad. Sci. Torino, II, vol. xix, pp. 85-118. *See* **Cornalia,
Emilio.**

Parfitt, Edward. The fauna of Devon. Part IX. Sessile-eyed Crustacea. [8vo, 25
pages.] "Reprinted from the Transactions of the Devonshire Association for the
Advancement of Science, Literature, and Art. 1873."

Pennant, Thomas. "The British Zoology, 4th Edit., 4 vols., with 279 plates, 4to.
London, 1777."

Pidgeon, Edward. The Classes Crustacea, etc. *See* **Griffith, Edward.**

Rathke, Heinrich. "Beitrag zur Fauna der Krimm. < Memoiren der kaiserlichen
Akademie der Wissenschaften zu St. Petersburg, Theil iii, pp. 291-454, 773-774.
1837."

Rathke, Heinrich. Beiträge zur Fauna Norwegens. < Nova Acta Academiæ
Cæsareæ Leopoldino-Carolinæ Naturæ Curiosorum, tom. xx, pp. 1-264c., taf. i-xii.
Breslau and Bonn, 1843.

Rathke, Jens. "Jagttagelser henhørende til Indvoldsormenes og Blöddyrenes naturhistorie; med anmärkningar af O. Fabricius. < Skrivter af naturhistorie-
Selskabet, vol. v, pp. 61-153, tab. ii-iii. Copenhagen, 1799."

Reinhardt, Johann T. Fortegnelse over Grønlands Krebsdyr, Annelider og Indvold-
sorme. < Naturhistorisk Bidrag til en Beskrivelse af Grønland, pp. 28-49.
"Særskilt Aftryk af Tillægene til 'Grønland, geographisk og statistisk beskrevnet'
af H. Rink." Copenhagen, 1857.

Risso, Antoine. Histoire naturelle des Crustacés des environs de Nice. [8vo, 176
pages, 3 plates.] Paris, 1816.

Risso, Antoine. "Histoire naturelle de l'Europe méridionale, tome v. Paris, 1826."

Ritzema Bos, Jan. *See* Bos, Jan Ritzema.

Roux, Jean Louis Florent Polydore. "Crustacés de la Méditerranée et de son
Littoral décrits et lithographiés. Marseilles, 1829-1830."

Saenger, Nicholas. "Preliminary account of an exploration of the fauna of the Baltic. <Communications of the Imp. Society of Nat. Sc. Anthropol. and Ethnol. of the Univers. of Moscow, vol. viii, pp. 22–34. 1869."

Samouelle, George. The Entomologist's useful Compendium; or an introduction to the knowledge of British Insects. [8vo, 496 pp., 12 plates.] London, 1819.

Sars, George Ossian. [Om en anomal Gruppe af Isopoder.] <Forhandlinger i Videnskabs-Selskabet i Christiania, Aar 1863, pp. 205–221. Christiania, 1864.

Sars, George Ossian. Beretning om en i Sommeren, 1865, foretagen Zoologisk Reise ved Kysterne af Christianias og Christiansands Stifter. [8vo, 47 pages.] <"Nyt Magazin for Naturvidenskaberne." Christiania, 1866.

Sars, George Ossian. Histoire naturelle des Crustacés d'eau douce de Norvège. 1me livraison. Les Malacostracés. [4to, 145 pages, 10 plates.] Christiania, 1867.

Sars, George Ossian. Undersøgelser over Christianiafjordens Dybvandsfauna anstillede paa en i Sommeren 1868 foretagen zoologisk Reise. [8vo, 58 pages.] <"Nyt Magazin for Naturvidenskaberne." Christiania, 1869.

Sars, George Ossian. Undersøgelser over Hardangerfjordens Fauna. <Forhandlinger i Videnskabs-Selskabet i Christiania, Aar 1871, pp. 246–286. Christiania, 1872.

Sars, George Ossian. Bidrag til Kundskaben om Dyrelivet paa vore Havbanker. <Forhandlinger i Videnskabs-Selskabet i Christiania, Aar 1872, pp. 73–119. Christiania, 1873.

Sars, George Ossian. Prodromus descriptionis Crustaceorum et Pycnogonidarum, quae in expeditione Norvegica Anno 1876, observavit G. O. Sars. <Archiv for Mathematik og Naturvidenskab, Bind ii, pp. 337*[237]–271. Christiania, 1877.

Sars, Michael. Oversigt over de i den norsk-arctiske Region forekommende Krebsdyr. <Forhandlinger i Videnskabs-Selskabet i Christiania, Aar 1858, pp. 122–163. Christiania, 1859.

Sars, Michael. [Beskrivelse af en ny Slægt og Art af Isopoder: Munnopsis typica Sars.] <Forhandlinger i Videnskabs-Selskabet i Christiania, Aar 1860, pp. 84–85. Christiania, 1861.

Sars, Michael. Bidrag til Kundskab om Christianiafjordens Fauna. [104 pages, 7 plates.] <"Nyt Magazin for Naturvidenskaberne." Christiania, 1868.

Sars, Michael. Fortsatte Bemærkninger over det dyriske Livs Udbredning i Havets Dybder. <Forhandlinger i Videnskabs-Selskabet i Christiania, Aar 1868, pp. 246–275. Christiania, 1869.

Savigny, Jules Cæsar. Description de l'Égypte ou Recueil des Observations et des Recherches pendant l'Expédition de l'Armée Française. Histoire naturelle, Planches, Zoologie, Crustacés. [13 folio plates.] Paris, 1817.

Say, Thomas. An account of the Crustacea of the United States. <Journal of the Academy of Natural Science, vol. i, part i, pp. 57–63, 65–80, pl. iv, pp. 97–101, 155–169, 1817; part ii, pp. 235–253, 313–319, 374–401, 423–441. Philadelphia, 1817–1818.

Say, Thomas. Observations on some of the animals described in the account of the Crustacea of the United States. <Journal of the Academy of Natural Science, vol. i, part ii, pp. 442–444. Philadelphia, 1818.

---

*In this volume the paging from 200 to 268 is incorrectly printed 300–368. The separata are paged 237–371.

Schiödte, Jörgen C. On the structure of the mouth in sucking Crustacea, part i, Cymothoæ. < Annals and Magazine of Natural History, IV, vol. i, pp. 1-25, pl. i, 1868.—Parts ii, Anthura, iii, Laphystius. < Ibid., vol. xviii, pp. 253-266 and 295-305. London, 1877.
  Translated from "Naturhistorisk Tidsskrift III, vol. iv with 2 plates, and vol. x with 5 plates. Copenhagen, 1866 and 1875."

Schiödte, Jörgen C. Sur la propagation et les metamorphoses des Crustacés suceurs de la famille des Cymothoadiens. < Comptes Rendus, tome lxxxvii, pp. 52-55. Paris, 1878.
  Translated in Annals and Magazine of Natural History, V, vol. ii, pp. 195-197. London, 1878.

Smith, Sidney Irving. This Report, part i, pp. 537-747. *See* **Verrill, Addison E.**

Smith, Sidney Irving *and* **Harger, Oscar.** Report on the dredgings in the region of St. George's Banks in 1872. < Transactions of the Connecticut Academy of Arts and Sciences, vol. iii, part i, pp. 1-57, pl. i-viii. New Haven, 1874.

Smith, Sidney Irving. The Crustacea of the fresh waters of the United States. <This Report, part ii, pp. 637-665, pl. i-iii. Washington, 1874. [Descriptions of Asellus and of Asellopsis by O. Harger].

Smith, Sidney Irving. The stalk-eyed Crustaceans of the Atlantic coast of North America north of Cape Cod. < Transactions of the Connecticut Academy, vol. v, pp. 27-136, pl. viii-xii. New Haven, 1879.

Smith, Sidney Irving. Proc. U. S. Nat. Mus., 1879, vol. ii, pp. 157, 158. *See* **Harger, Oscar.**

Smith, Sidney Irving. Occurrence of Chelura terebrans, a Crustacean destructive to the timber of submarine structures, on the coast of the United States. < Proceedings of the United States National Museum, 1879, vol. ii, pp. 232-235. Washington, 1880.

Stalio, Luigi. Catalogo Metodico e Descrittivo dei Crostacei Podottalmi ed Edriottalmi dell'Adriatico. [8vo, 274 pages.] "(Estr. dal vol. iii, serie v, degli Atti dell' Instituto Stesso.)" Venice, 1877.

Sowerby, James. The British Miscellany: or coloured figures of new, rare, or little known animal subjects; many not before ascertained to be inhabitants of the British Isles. [2 vols. in one, 8vo, 76 plates.] London, 1804-1806.

Stebbing, Thomas Roscoe Rede. A Sphæromid from Australia, and Arcturidæ from South Africa. <Annals and Magazine of Natural History, IV, vol. xii, pp. 95-98, pl. iii A. London, 1873.

Stebbing, Thomas Roscoe Rede. On a new species of Arcturus (A. damnoniensis). <Annals and Magazine of Natural History, IV, vol. xiii, pp. 291-292, pl. xv. London, 1874.

Stebbing, Thomas Roscoe Rede. A new Australian Sphæromid, Cyclura venosa; and notes on Dynamene rubra and viridis. < Journal of the Linnean Society, Zoology, vol. xii, pp. 146-151, pl. vi-vii. London, 1874.

Stebbing, Thomas Roscoe Rede. The sessile-eyed Crustacea of Devon. [8vo, 10 pages, 1 plate.] "Reprinted from the Transactions of the Devonshire Association for the Advancement of Science, Literature, and Art. 1874."

Stebbing, Thomas Roscoe Rede. On some new exotic sessile-eyed Crustaceans. <Annals and Magazine of Natural History, IV, vol. xv, pp. 184-188, pl. xv A. London, 1875.

Stebbing, Thomas Roscoe Rede. Description of a new species of sessile-eyed Crustacean and other notices. <Annals and Magazine of Natural History, IV, vol. xvii, pp. 73–80, pl. iv–v. London, 1876.

Stebbing, Thomas Roscoe Rede. Notes on sessile-eyed Crustaceans with description of a new species. <Annals and Magazine of Natural History, V, vol i, pp. 31–37, pl. v. London, 1878.

Stebbing, Thomas Roscoe Rede. Sessile-eyed Crustacea of Devonshire. Supplementary list. [8vo, 9 pages.] "Reprinted from the Transactions of the Devonshire Association for the Advancement of Science, Literature, and Art." 1879.

Steenstrup, Japetus, *and* **Lütken, Christian Friedrich.** Mindre Meddelelser fra Kjöbenhavns Universitets zoologiske Museum.—2. Forelöbig Notits om danske Hav-Krebsdyr. <Videnskabelige Meddelelser fra den Naturhistoriske Forening i Kjöbenhavn, 1861, II, vol. iii, pp. 274–276. Copenhagen, 1862.

Stimpson, William. Synopsis of the Marine Invertebrata of Grand Manan; or the region about the mouth of the Bay of Fundy, New Brunswick. [4to, 66 pp., 3 plates.] <Smithsonian Contributions to Knowledge, vol. vi. Washington, 1853.

Stimpson, William. Descriptions of some new marine Invertebrata. By William Stimpson, Zoologist to the U. S. Surveying Expedition to the North Pacific, Japan Seas, etc., under direction of Commander C. Ringgold, U. S. N. <Proceedings of the Academy of Natural Science, vol. vii, 1855, pp. 385–394. Philadelphia, 1855.

Stimpson, William. On an oceanic Isopod found near the southeastern shores of Massachusetts. <Proceedings of the Academy of Natural Science, vol. xiv, 1862, pp. 133–134. Philadelphia, 1862.

Stimpson, William. Synopsis of the marine Invertebrata collected by the late Arctic Expedition under Dr. I. I. Hayes. <Proceedings of the Academy of Natural Science, vol. xv, 1863, pp. 138–142. Philadelphia, 1863.

Streets, Thomas Hale, *and* **Kingsley, John Sterling.** An examination of types of some recently described Crustacea. <Bulletin of the Essex Institute, vol. ix, pp. 103–108. Salem, 1877.

Templeton, Robert. Description of a minute crustaceous animal from the Island of Mauritius. <Transactions of the Entomological Society, vol. ii, pp. 203–207, pl. xviii. London, "1836."

Templeton, Robert. Catalogue of Irish Crustacea Myriapoda and Arachnöidea, selected from the papers of the late John Templeton, Esq. <Loudon's Magazine of Natural History, vol. ix, pp. 9–14. London, 1836.

Thompson, William. On the Teredo navalis and Limnoria terebrans as at present existing in certain localities on the coasts of the British Islands. <Edinburgh New Philosophical Journal, vol. xviii, pp. 121–130. 1835.

Thompson, William. Note on the Teredo norvegica, Xylophaga dorsalis, Limnoria terebrans and Chelura terebrans combined in destroying the submerged wood-work at the Harbor of Ardrossan, on the coast of Ayrshire. <Annals and Magazine of Natural History, vol. xx, pp. 157–164. London, 1847.

Thompson, William. Additions to the fauna of Ireland, Crustacea. <Annals and Magazine of Natural History, vol. xx, pp. 237–250. London, 1847.

Van Beneden, Pierre Joseph. Recherches sur la Faune littorale de Belgique. Crustacés. [4to, 180 pages, 32 plates.] "Extrait du tome xxxiii des Mémoires de l'Académie royale de Belgique." Bruxelles, 1861.

Verrill, Addison Emory. On the distribution of marine animals on the southern coast of New England. < American Journal of Science and Arts, III, vol. ii, pp. 357-362. New Haven, 1871.

Verrill, Addison Emory. Results of recent dredging expeditions on the coast of New England. (No. 1). < American Journal of Science and Arts, III, vol. v, pp. 1-16, Jan. 1873.—(No. 2). < Ibid., pp. 98-106, Feb. 1873.—No. 3. < Ibid., vol. vi, pp. 435-441, Dec. 1873.—No. 4. < Ibid., vol. vii, pp. 38-46, Jan. 1874.—No. 5. < Ibid., pp. 131-138, Feb. 1874.—No. 6. < Ibid..pp. 405-414, pl. iv-v, Apr., 1874.— No. 7. < Ibid., pp. 498-505, pl. vi-viii, May, 1874. New Haven, 1873-4.

Verrill, Addison Emory. Explorations of Casco Bay by the United States Fish Commission in 1873. < Proceedings of the American Association for the Advancement of Science, Portland Meeting, 1873, pp. 340-395, pl. i-vi. Salem, 1874.

Verrill, Addison Emory. Report upon the invertebrate animals of Vineyard Sound and the adjacent waters, with an account of the physical characters of the region. < This Report, part i, pp. 295-778 (1-478), pl. i-xxxviii. Washington, 1874.
Published also separately with the above title or, Invertebrata of Southern New England, by A. E. Verrill and S. I. Smith. [8vo, 478 pages, 38 plates.] Washington, 1874.

Verrill, Addison Emory, Smith, Sidney Irving, and **Harger, Oscar.** Catalogue of the marine invertebrate animals of the Southern Coast of New England and adjacent waters. < This Report, part i, pp. 537-747 (243-453), pl. i-xxxviii. Washington, 1874.
Published also as a part of the above Report upon the invertebrate animals of Vineyard Sound and adjacent waters or Invertebrata of Southern New England, by A. E. Verrill and S. I. Smith, pp. 243-453. Washington, 1874.

Verrill, Addison Emory. Results of dredging expeditions off the New England Coast in 1874. < American Journal of Science and Arts, III, vol. ix, pp. 411-415, vol. x, pp. 36-43, pl. iii-iv, and pp. 196-202. New Haven, 1875.

Wagner, Nicholas. Recherches sur le système circulatoire et les organes de respiration chez le Porcellion élargi (Porcellio dilatatus Brandt). < Annales des Sciences naturelles, Zoologie, V, tome iv, pp. 317-322, pl. xiv B. Paris, 1865.

Wagner, Nicholas. Observations sur l'organisation et le développement des Ancees. < Bulletin de l'Académie Impériale des Sciences de St. Pétersbourg, tome x, pp. 497-502. 1866.

Westwood, John Obadiah. Extrait des recherches sur les Crustacés du genre Pranize de Leach. < Annales des Sciences naturelles, tome xxvii, pp. 316-322, pl. vi. Paris, 1832.

Westwood, John Obadiah. British Sessile-eyed Crustacea. *See* **Bate, C. Spence.**

White, Adam. List of the specimens of Crustacea in the collection of the British Museum. [143 pages.] London, 1847.

White, Adam. List of the specimens of British animals in the collection of the British Museum, part iv, Crustacea. [141 pages.] London, 1850.

White, Adam. A popular history of British Crustacea, comprising a familiar account of their classification and habits. [358 pages, 20 plates.] London, 1857.

Whiteaves, Joseph Frederick. Notes on a deep-sea dredging-expedition round the Island of Anticosti, in the Gulf of St. Lawrence. < Annals and Magazine of Natural History, IV, vol. x, pp. 341-354. London, 1872.

Whiteaves, Joseph Frederick. Report of a second deep-sea dredging expedition [in 1872] to the Gulf of St. Lawrence, with some remarks on marine fisheries of the Province of Quebec. [8vo, 22 pages.] Montreal, 1873.

Whiteaves, Joseph Frederick. On recent deep-sea dredging operations in the Gulf of St. Lawrence. < American Journal of Science and Arts, III, vol. vii, pp. 210–219. New Haven, 1874.

Whiteaves, Joseph Frederick. Report on further deep-sea dredging operations in the Gulf of St. Lawrence [in 1873], with notes on the present condition of the marine fisheries and oyster-beds of part of that region. [8vo, 29 pages.] "Ottawa, 1874."

Willemoes-Suhm, Rudolf von. Von der Challenger Expedition. Briefe an C. Th. E. v. Siebold. II, Sidney, im April, 1874. < Zeitschrift für wissenschaftliche Zoologie, Band xxiv, pp. ix–xxiii. Leipzig, 1874.

Willemoes-Suhm, Rudolf von. On some Atlantic Crustacea from the 'Challenger' Expedition. < Transactions of the Linnean Society of London, II, Zoology, vol. i, pp. 23–59, pl. vi–xiii. London, 1875.

Willemoes-Suhm, Rudolf von. Preliminary report to Professor Wyville Thomson, F. R. S. Director of the civilian scientific staff, on observations made during the earlier part of the voyage of H. M. S. 'Challenger.' < Proceedings of the Royal Society, vol. xxiv, pp. 569–585.—On Crustacea observed during the cruise of H. M. S. 'Challenger' in the Southern Sea. < Tom. cit., pp. 585–592. London, 1876.

Woodward, Henry. Crustacea. < Encyclopædia Britannica, 9th edition, vol. vi, pp. 632–666. Edinburgh and Boston, 1877.

Zaddach, Ernest Gustav. Synopseos Crustaceorum Prussicorum Prodromus. [4to, 39 pages.] Regiomonti (Königsberg), "1844."

TABLE OF CONTENTS.

XIV.—Report on the Marine Isopoda of New England and adjacent waters, by Oscar Harger	297
Isopoda	297
Synoptical table of families	304
Oniscidæ	305
Bopyridæ	311
Asellidæ	312
Munnopsidæ	328
Idoteidæ	335
Arcturidæ	361
Sphæromidæ	367
Limnoriidæ	371
Cirolanidæ	376
Ægidæ	382
Cymothoidæ	390
Anthuridæ	396
Gnathiidæ	408
Tanaidæ	413
Geographical distribution	428
List of authorities	436
Table of contents	451
Explanation of the plates	453
Alphabetical index	459

EXPLANATION OF THE PLATES.

PLATE I.

FIGURE 1.—Philoscia vittata Say (p. 306); dorsal view, enlarged **six diameters**; natural size indicated by cross at the right.
2.—Scyphacella arenicola Smith (p. 307); dorsal view, enlarged about twelve diameters; natural size indicated by cross at the right.
3.—Actoniscus ellipticus Harger (p. 309); dorsal view, enlarged ten diameters; natural size indicated by line at the right.
4.—Jæra albifrons Leach (p. 315); female; dorsal view, enlarged about ten diameters.
5.—The same; maxilliped from the left side, exterior view, enlarged twenty-five diameters; *p*, palpus; *l*, external lamella.
6.—The same; maxillæ, enlarged twenty-five diameters; *a*, outer, or second, **pair of maxillæ**; *b*, inner, or first, pair of maxillæ; *i*, inner, *e*, outer lobe.
7.—The same; inferior surface of the pleon of a female.
8.—The same; inferior surface of the pleon of a male.

(All the figures were drawn from nature by O. Harger.)

PLATE II.

FIGURE 9.—Janira alta Harger (p. 321); dorsal view, **enlarged five diameters**; natural size indicated by line at the right.
10.—Janira spinosa Harger (p. 323); dorsal view **of** female, enlarged six diameters.
11.—Munnopsis typica M. Sars (p. 330); dorsal view of male, enlarged about two diameters; *b*, maxilliped; *a*, basal segment; *l*, external lamella; 2 and 3, second and third segments of palpus of maxilliped; *c*, outer maxillæ; *d*, inner maxillæ; *e*, one of the second pair of legs of the male; *f*, one of the natatory legs; *p*, abdominal operculum of the female, external view.

(Figures 9 and 10 were drawn from nature by O. Harger; figure 11 is copied from M. Sars, drawn by G. O. Sars.)

PLATE III.

FIGURE 12.—Janira alta (p. 321); *a*, maxilliped; *p*, palpus of maxilliped; *l*, external lamella; *b*, mandibles; *P*, palpus of mandible; *d*, denticigerous lamella; *m*, molar process, enlarged twenty-five diameters.
13.—The same; inferior surface of the pleon, *a* in the female, *b* in the male, enlarged ten diameters; *a*, single opercular plate in the female; *b*, external; *c*, median plate of operculum of male.
14.—Munna Fabricii Krö.. (p. 318); female; dorsal view, enlarged about **twenty** diameters; natural size indicated by line at the right.
15.—**Eurycope robusta** Harger (p. 332); female; dorsal view, enlarged six diameters; natural size indicated by line at the right; *a*, antennula, enlarged twenty diameters; *b*, maxilliped; *c*, mandible; *d*, one of the first pair of legs, each enlarged twenty diameters; *d*, propodus and dactylus of the first pair of legs, enlarged about thirty-eight diameters; *e*, **propodus** and dactylus of the second pair of legs, enlarged twenty

diameters; *f*, one of the sixth pair of legs; *g*, uropod, each enlarged twenty diameters.

(Figure 14 was drawn from nature by Mr. J. H. Emerton, the others by O. Harger.)

PLATE IV.

FIGURE 16.—Chiridotea cœca Harger (p. 338); dorsal view, enlarged nearly four diameters; natural size indicated by the line at the right.

17.—The same; *a*, antennula; *b*, antenna; each enlarged twelve diameters.

18.—The same; *a*, maxilliped from the right side, external view; *l*, external lamella; *m*, maxilliped proper; 1, 2, 3, first, second, and third segments of the palpus of the maxilliped, enlarged twenty diameters; *b*, one of the first pair of legs, magnified twelve diameters; *c*, uropod from the left side, inner view, showing the two rami articulated near the tip.

19.—The same; pleopods of second pair from the right side, anterior views, enlarged ten diameters; *a*, common form in males; *b*, rarer form in male; *s*, elongated stylet, articulated near the base of the inner lamella; *c*, form in the female.

20.—Chiridotea Tuftsii Harger (p. 340); female; dorsal view, enlarged five diameters; natural size indicated by the line at the right.

21.—The same; left maxilliped, enlarged twenty-five diameters; *e*, external lamella; *m*, basal segment; 1, 2, 3, segments of palpus.

22.—The same; pleopod of the second pair, from a male, enlarged twenty diameters; *s*, elongated stylet, articulated near the base of the inner lamella.

(All the figures were drawn from nature by O. Harger.)

PLATE V.

FIGURE 23.—Chiridotea Tuftsii Harger (p. 340); *a*, antennula; *b*, antenna; *c*, leg of the first pair; *d*, leg of the fourth pair; all enlarged twelve diameters; *e*, left uropod, or opercular valve, inner view, enlarged ten diameters.

24.—Idotea irrorata Edwards (p. 343); dorsal view, enlarged two diameters; natural size shown by the line on the left.

25.—The same; *a*, antennula; *b*, antenna; *e*, left uropod or opercular valve, external view; all enlarged six diameters.

26.—The same; *a*, right maxilliped, enlarged twelve diameters, *l*, external lamella; *m*, basal segment; 1, 2, 3, 4, segments of palpus of maxilliped; *b*, pleopod of the second pair from a male, enlarged eight diameters, showing stylet, *s*, articulated near the base of the inner lamella.

27.—Idotea phosphorea Harger (p. 347); dorsal view, enlarged about two diameters; natural size shown by the line on the right.

28.—The same; *a*, antenna, enlarged six diameters; *b*, maxilliped, enlarged twelve diameters, showing, *l*, external lamella; *m*, basal segments; 1, 2, 3, 4, segments of the palpus of maxilliped; *c*, leg of the first pair; *d*, leg of the second pair, both enlarged six diameters; *e*, right uropod, or opercular valve, inner view, enlarged six diameters.

29.—The same; pleopod of the second pair from a male, enlarged eight diameters; *s*, stylet articulated near the base of the inner lamella; *s′*, distal end of stylet reversed and enlarged thirty diameters.

(Figure 24 was drawn by Mr. J. H. Emerton, the others by O. Harger.)

PLATE VI.

FIGURE 30.—Idotea robusta Krøyer (p. 349); dorsal view, enlarged two diameters; natural size shown by the line at the right.

31.—The same; *a*, antenna; *b*, leg of the first pair, each enlarged six diameters; *e*, left uropod, or opercular valve, inner view, enlarged four diameters.

EXPLANATION OF THE PLATES. 455

FIGURE 32.—The same; *a*, maxilliped, enlarged twelve diameters; *l*, external lamella; 1, 2, 3, 4, segments of palpus; *b*, maxilla of the outer or second pair; *c*, pleopod of the second pair from a male, enlarged six diameters; *s*, stylet articulated near the base of the inner lamella.

33.—Synidotea nodulosa Harger (p. 354); dorsal view, enlarged four diameters; natural size indicated by the line at the right.

34.—The same; *a*, antennula; *f*, flagellar segment; *b*, antenna; *c*, leg of the first pair from the right side; *d*, right uropod, or opercular valve, all enlarged ten diameters.

35.—The same; *a*, maxilliped from the right side, showing, *l*, external lamella; *m*, basal segment; 1, 2, 3, segments of palpus, enlarged twenty diameters; *b*, maxilla of the outer or second pair; *c*, maxilla of the inner or first pair, both enlarged twenty diameters; *d*, pleopod of the second pair from a male, enlarged twelve diameters; *s*, stylet articulated near the base of the inner lamella.

36.—Erichsonia attenuata Harger (p. 356); dorsal view, enlarged three diameters, natural size indicated by the line at the right.

(Figures 30 and 36 were drawn by Mr. J. H. Emerton, the others by O. Harger.)

PLATE VII.

FIGURE 37.—Erichsonia attenuata Harger (p. 356); *a*, antennula; *b*, antenna, each enlarged twelve diameters; *c*, maxilliped, showing, *l*, external lamella, enlarged thirty diameters; *d*, uropod, or opercular valve, enlarged twelve diameters; *e*, pleopod of the second pair from a male, enlarged fifteen diameters; *s*, stylet, articulated near the base of the inner lamella; *s'*, distal end of stylet, enlarged fifty diameters.

38.—Erichsonia filiformis Harger (p. 355); dorsal view, enlarged five diameters, natural size indicated by the line at the right.

39.—The same; *a*, antennula; *b*, antenna; *c*, leg of the first pair; *d*, uropod, or opercular valve, each enlarged twelve diameters.

40.—The same; *a*, maxilla of outer or second pair; *b*, maxilla of inner or first pair; *c*, mandible, showing molar process, *m*, and dentigerous lamella, *d*, all enlarged thirty diameters.

41.—The same; *a*, maxilliped, showing, *l*, external lamella; *m*, basal segment, and 1, 2, 3, 4, segments of palpus, enlarged thirty diameters; *b*, pleopod of the second pair from a male, enlarged fifteen diameters; *s*, stylet, articulated near the base of the inner lamella; *s'*, distal end of stylet, enlarged fifty diameters.

42.—Epelys trilobus Smith (p. 358); dorsal view, enlarged ten diameters; natural size indicated by the line at the right.

43.—The same; *a*, maxilliped from the left side, enlarged twenty diameters; *l*, external lamella; *m*, basal segment; 1, 2, 3, segments of palpus of maxilliped; *b*, pleopod of second pair from a male, enlarged twenty diameters; *s*, stylet, articulated near the base of the inner lamella; *s'*, end of stylet, enlarged fifty diameters.

(All the figures were drawn from nature by O. Harger.)

PLATE VIII.

FIGURE 44.—Epelys montosus Harger (p. 359); dorsal view, enlarged six diameters, natural size indicated by the line at the right.

45.—The same; *a*, antennula; *f*, flagellar segment; *b*, antenna; *c*, maxilliped from the left side; *l*, external lamella; *m*, basal segment; 1, 2, 3, segments of palpus; all the figures enlarged twenty diameters.

46.—The same; *a*, leg of the first pair, enlarged twenty diameters; *b*, right uropod or opercular valve, enlarged fifteen diameters.

456   REPORT OF COMMISSIONER OF FISH AND FISHERIES.

FIGURE 47.—The same; pleopod of the second pair, from a male, enlarged twenty diameters; *s*, stylet, articulated near the base of the inner lamella; *s'*, distal end of stylet, enlarged sixty-six diameters.
48.—Astacilla granulata Harger (p. 364); female; dorsal view, enlarged four diameters, natural size indicated by the line at the right; *a*, antennula of male; *b*, fourth thoracic segment of male; *c*, inferior surface of pleon of a male, showing opercular valves; all the figures enlarged four diameters.
49.—The same; *a*, flagellum of antenna, enlarged twenty diameters; *a'*, portion of inner margin of the same, enlarged one hundred diameters; *b*, one of the first pair of legs, upper surface, enlarged twenty diameters.
50.—The same; one of the fourth pair of legs, enlarged twenty diameters.
51.—The same; inner surface of left opercular plate, or uropod, from a female, enlarged twenty diameters.

(All the figures were drawn from nature by O. Harger.)

PLATE IX.

FIGURE 52.—Astacilla granulata Harger (p. 364); *a*, maxilliped; *m*, basal segment; *l*, external lamella; *b*, outer maxilla; *c*, inner maxilla; all enlarged twenty diameters.
53.—Sphaeroma quadridentatum Say (p. 368); dorsal view, enlarged five diameters; natural size indicated by the line at the right.
54.—The same; *a*, antennula; *b*, antenna; *c*, pleopod of the second pair, from a male, showing stylet, *s*, articulated near the base of the inner lamella; all the figures enlarged ten diameters.
55.—Limnoria lignorum White (p. 373); dorsal view, enlarged ten diameters; natural size indicated by the line at the right.
56.—The same; *a*, antennula; *b*, antenna; *c*, maxilliped; *d*, maxilla of the outer or second pair; *e*, maxilla of the inner or first pair; *f*, mandible, all enlarged twenty-five diameters; *e'*, distal end of outer lobe of first pair of maxillae, enlarged sixty-six diameters.
57.—The same; *a*, last segment of pleon, with attached uropods; dorsal view, enlarged ten diameters; *b*, uropod with dotted adjacent outline of last segment of pleon, enlarged thirty diameters; *c*, first pair of pleopods; *d*, pleopod of the second pair, from a male, showing stylet, *s*, articulated to the inner lamella; both figures enlarged twenty diameters.
58.—Cirolana concharum Harger, (p. 378); lateral view, enlarged about three diameters.

(Figure 53 was drawn by Mr. J. H. Emerton, 55 by Prof. S. I. Smith, 58 by Mr. J. H. Blake, and the others by O. Harger.)

PLATE X.

FIGURE 59.—Cirolana concharum Harger (p. 378); dorsal view, enlarged about three diameters. The natural size is shown by the line at the right.
60.—The same; antennula, enlarged ten diameters.
61.—The same; *a*, antenna enlarged ten diameters; *b*, maxilla of the outer or second pair; *c*, maxilla of the inner or first pair; *d*, mandible from the right side, inner view; *p*, palpus; *m*, molar area; the last three figures enlarged five diameters.
62.—The same; *a*, maxilliped from the right side, exterior view, showing, *l*, external lamella; *m*, basal segment; 1, 2, 3, 4, 5, segments of the palpus; *b*, leg of the fourth pair; both the figures enlarged five diameters.
63.—The same; uropod from the right side; inferior view, enlarged five diameters.
64.—Ægapsora Kröyer (p. 384); *a*, dorsal and *b* ventral views of a young individual. The central line indicates the length of the specimen, natural

EXPLANATION OF THE PLATES. 457

size, which is here enlarged three diameters. Adults attain about the size of the figure.

FIGURE 65.—Nerocila munda Harger (p. 392); dorsal view of the type specimen, enlarged about four diameters. The natural size is shown by the cross on the right; *a*, uropod, enlarged six diameters.

66.—Ægathoa loliginea Harger (p. 393); type specimen; *a*, dorsal, and *b*, ventral view, enlarged four diameters. Its natural size is shown by the line between the figures.

(Figure 59 was drawn by Mr. J. H. Blake, the others by O. Harger.)

PLATE XI.

FIGURE 67.—Livoneca ovalis White (p. 395); *a*, antennula; *b*, antenna; *c*, mandibular palpus; each enlarged twenty diameters; *d*, one of the first pair of legs; *e*, one of the seventh pair of legs; *f*, uropod; each enlarged ten diameters.

68.—Anthura polita Stimpson (p. 398); dorsal view, enlarged four diameters. The natural size is shown by the line at the right; *a*, antennula; *b*, antenna, each enlarged ten diameters; *c*, leg of the first pair; *d*, leg of the third pair; *e*, right pleopod of the first pair, interior view, showing inner ramus without cilia; *f*, pleopod of the second pair from a male, showing stylet articulated to inner lamella; each of the figures *c* to *f* enlarged eight diameters; *g*, lateral view of pleon, enlarged six diameters.

69.—The same; *a*, maxilliped, enlarged twenty diameters; *b*, maxilla, enlarged twenty-five diameters; *b'*, distal end of the same, enlarged sixty diameters.

70.—Paranthura brachiata Harger (p. 402); dorsal view, enlarged about three diameters; natural size shown by the line at the right; *a*, antennula; *b*, antenna, enlarged eight diameters; *c*, right maxilliped, enlarged sixteen diameters; *d*, maxilla, enlarged sixteen diameters; *d'*, distal end of the same, enlarged fifty diameters; *e*, leg of the first pair; *f*, first pleopod from the right side, inner view, showing ciliated inner lamella; *g*, pleopod of the second pair from a male, showing stylet articulated to the inner lamella; figures *e* to *g* enlarged eight diameters.

71.—Ptilanthura tenuis Harger (p. 406); male; dorsal view, enlarged about four diameters; *a*, inferior view of the head and first thoracic segment, enlarged eight diameters; the flagellum of the antennulæ omitted; *b*, maxilliped; *c*, maxilla, each enlarged fifty diameters; *d*, first right pleopod, seen from within, showing ciliated inner lamella; *e*, second left pleopod, showing stylet *s* articulated to the inner lamella in the males.

72.—The same; one of the first pair of legs of a male, enlarged sixteen diameters.

73.—The same; female; dorsal view of the head, enlarged twenty-five diameters.

(Figure 71, excepting *b-d*, was drawn by Mr. J. H. Emerton, the others by O. Harger.)

PLATE XII.

FIGURE 74.—Ptilanthura tenuis Harger (p. 406); *a*, antennula; *b*, antenna; each enlarged twenty diameters, from a male.

75.—Gnathia cerina Harger (p. 410); male; dorsal view, enlarged ten diameters.

76.—The same; *a*, antennula; *b*, antenna, each enlarged thirty-eight diameters; *c*, mandibles (*l*, left, *r*, right), enlarged thirty-eight diameters; *d*, first leg or first gnathopod from the right side, enlarged twenty-five diameters; all the figures from the male sex.

77.—The same (p. 411); female; dorsal view, enlarged ten diameters.

FIGURE 78.—The same; *a*, one of the first pair of legs or first gnathopod of a female, enlarged thirty-eight diameters; *b*, one of the first pair of legs in a young, parasitic individual, enlarged sixty diameters; *c*, pleon, with the last and part of the penultimate thoracic segments of a female, dorsal view, enlarged twenty diameters; *d*, pleopod of a young, parasitic individual, enlarged sixty diameters; *e*, pleopod of an adult male, enlarged sixty diameters.
79.—The same; young male; dorsal view, enlarged twenty diameters.
80.—Leptochelia algicola Harger (p. 421); male; lateral view, enlarged twenty diameters; natural size indicated by the line above.

(All the figures were drawn from nature by O. Harger.)

PLATE XIII.

FIGURE 81.—Tanais vittatus Lilljeborg (p. 418); dorsal view, enlarged eight diameters. The transverse bands of hairs on the pleon are not sufficiently distinct.
82.—The same; one of the first pair of pleopods, enlarged thirty diameters.
83.—Leptochelia algicola Harger (p. 421); female; dorsal view, enlarged twenty diameters; natural size indicated by the line at the right.
84.—The same; *a*, antennula; *b*, one of the first pair of legs; both from a female specimen and enlarged twenty-five diameters.
85.—The same; hand, or propodus and dactylus of the first pair of legs, enlarged forty-eight diameters, showing the comb of setæ on the propodus.
86.—The same; uropods of a male, enlarged seventy diameters; *b*, basal segment; *i*, inner six-jointed ramus; *o*, outer ramus.
87.—Leptochelia limicola Harger (p. 424); female; dorsal view, enlarged twenty diameters; natural size shown by the line at the right.
88.—The same; *a*, antennula; *b*, antenna; *c*, leg of the first pair; *d*, leg of the second pair; all from the female sex and enlarged twenty-five diameters.
89.—Leptochelia rapax Harger (p. 424); male; dorsal view, enlarged about twelve diameters.
90.—The same; hand, or propodus and dactylus of male, enlarged sixteen diameters.
91.—Leptochelia cœca Harger (p. 427); type specimen, female; *a*, antennula; *b*, leg of the first pair; *c*, uropod; each enlarged fifty diameters.

(All the figures were drawn from nature by O. Harger.)

PLATE I.

FIGURE 1.—Philoscia vittata Say (p. 306); dorsal view, enlarged six diameters; natural size indicated by cross at the right.

2.—Scyphacella arenicola Smith (p. 307); dorsal view, enlarged about twelve diameters; natural size indicated by cross at the right.

3.—Actoniscus ellipticus Harger (p. 309); dorsal view, enlarged ten diameters; natural size indicated by line at the right.

4.—Jæra albifrons Leach (p. 315); female; dorsal view, enlarged about ten diameters.

5.—The same; maxilliped from the left side, exterior view, enlarged twenty-five diameters; p, palpus; l, external lamella.

6.—The same; maxillæ, enlarged twenty-five diameters; a, outer, or second, pair of maxillæ; b, inner, or first, pair of maxillæ; i, inner, e, outer lobe.

7.—The same; inferior surface of the pleon of a female.

8.—The same; inferior surface of the pleon of a male.

(All the figures were drawn from nature by O. Harger.)

Report U. S. F. C. 1878.—Harger. Marine Isopods. **Plate I.**

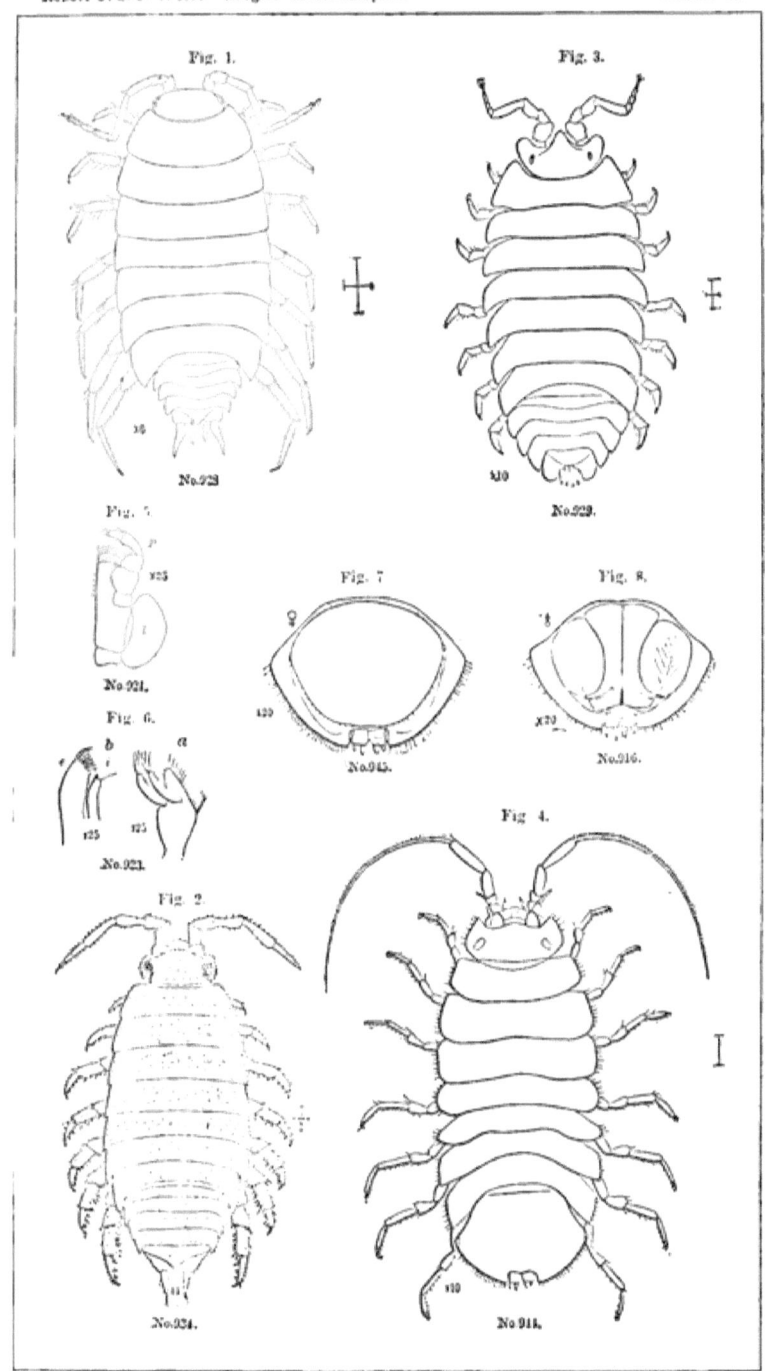

PLATE II.

FIGURE 9.—Janira alta Harger (p. 321); dorsal view, enlarged five diameters; natural size indicated by line at the right.

10.—Janira spinosa Harger (p. 323); dorsal view of female, enlarged six diameters.

11.—Munnopsis typica M. Sars (p. 330); dorsal view of male, enlarged about two diameters; *b*, maxillipeds; *a*, basal segment; *l*, external lamella; 2 and 3, second and third segments of palpus of maxillipeds; *c*, outer maxillæ; *d*, inner maxillæ; *e*, one of the second pair of legs of the male; *f*, one of the natatory legs; *g*, abdominal operculum of the female, external view.

(Figures 9 and 10 were drawn from nature by O. Harger; figure 11 is copied from M. Sars, drawn by G. O. Sars.)

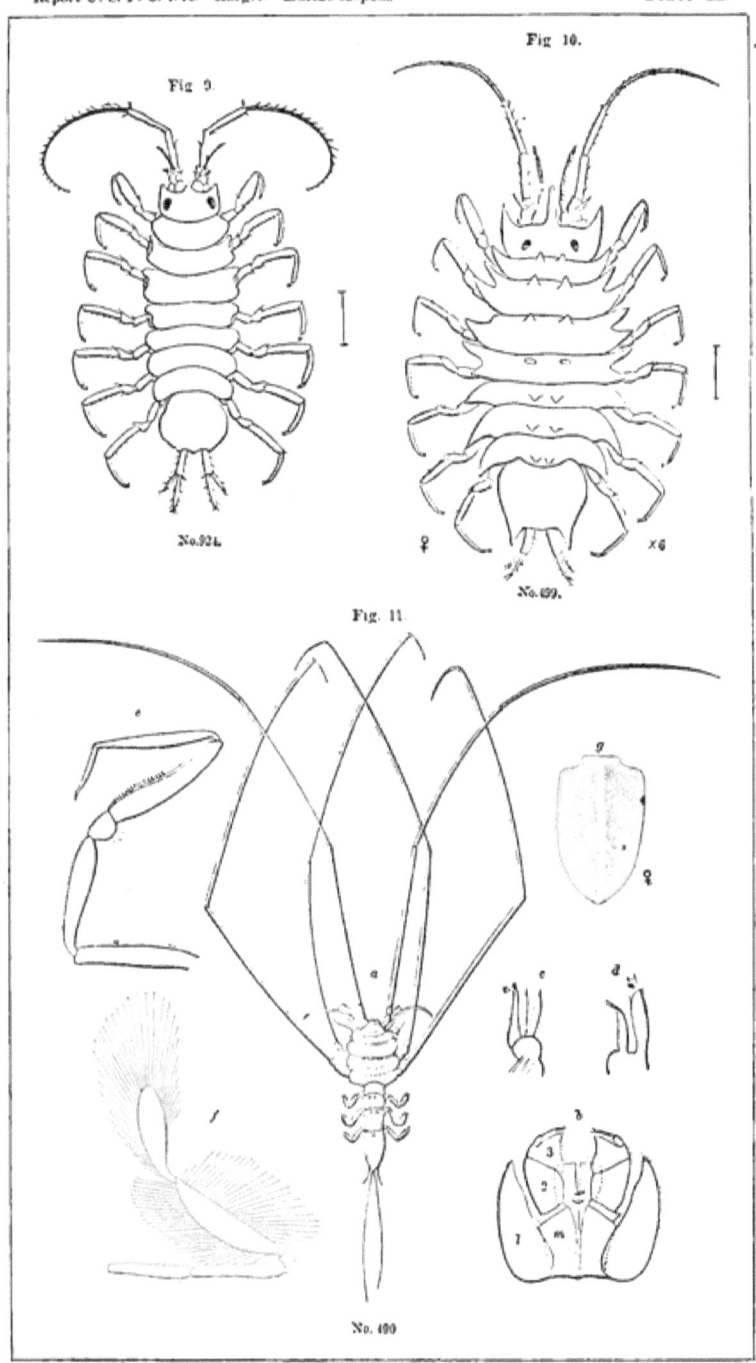

PLATE III.

FIGURE 12.—Janira alta (p. 321); *a*, maxilliped; P, palpus of maxilliped; *l*, external lamella; *b*, mandible; *P*, palpus of mandible; *d*, dentigerous lamella; *m*, molar process, enlarged twenty-five diameters.

13.—The same; inferior surface of the pleon, *a* in the female, *b* in the male, enlarged ten diameters; *a*, single opercular plate in the female; *b*, external; *c*, median plate of operculum of male.

14.—Munna Fabricii Kröyer (p. 325); female; dorsal view, enlarged about twenty diameters; natural size indicated by line at the right.

15.—Eurycope robusta Harger (p. 332); female; dorsal view, enlarged six diameters; natural size indicated by line at the right; *a*, antennula, enlarged twenty diameters; *b*, maxilliped; *c*, mandible; *d*, one of the first pair of legs, each enlarged twenty diameters; *d'*, propodus and dactylus of the first pair of legs, enlarged about thirty-eight diameters; *e*, propodus and dactylus of the second pair of legs, enlarged twenty diameters; *f*, one of the sixth pair of legs; *g*, uropod, each enlarged twenty diameters.

(Figure 14 was drawn from nature by Mr. J. H. Emerton, the others by O. Harger.)

Plate III.

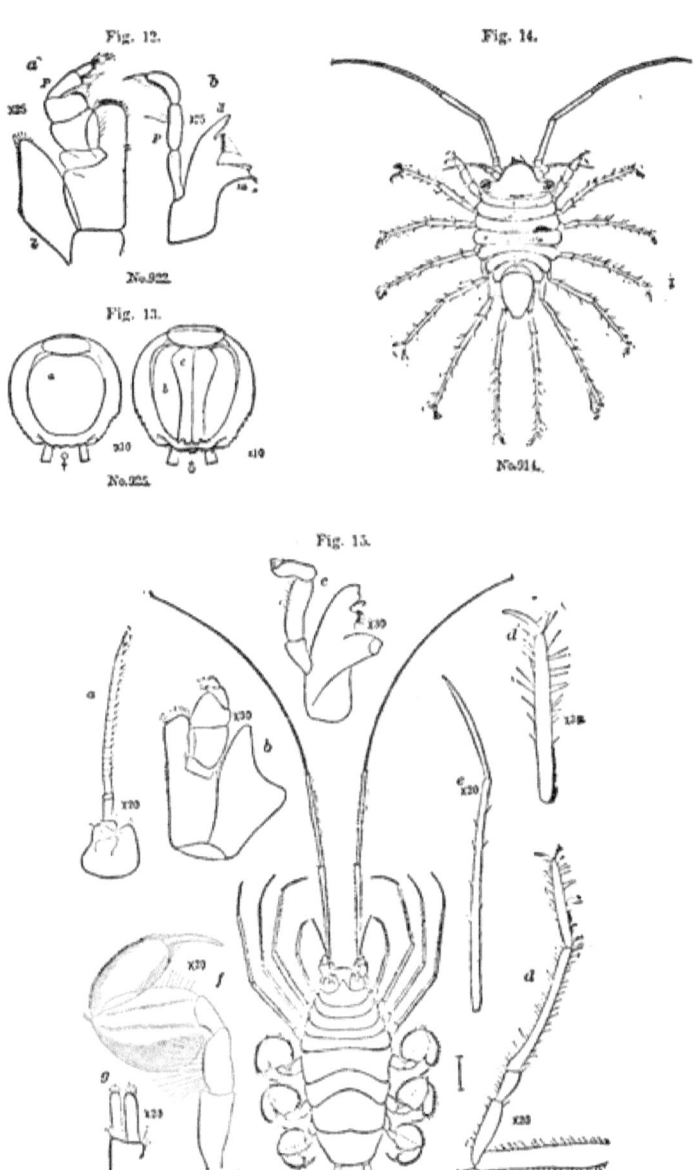

PLATE IV.

FIGURE 16.—Chiridotea cœca Harger (p. 338); dorsal view, enlarged nearly four diameters; natural size indicated by the line at the right.

17.—The same; *a*, antennula; *b*, antenna; each enlarged twelve diameters.

18.—The same; *a*, maxilliped from the right side, external view; *l*, external lamella; *m*, maxilliped proper; 1, 2, 3, first, second, and third segments of the palpus of the maxilliped, enlarged twenty diameters; *b*, one of the first pair of legs, magnified twelve diameters; *c*, uropod from the left side, inner view, showing the two rami articulated near the tip.

19.—The same; pleopods of second pair from the right side, anterior views, enlarged ten diameters; *a*, common form in males; *b*, rarer form in male; *s*, elongated stylet, articulated near the base of the inner lamella; *c*, form in the female.

20.—Chiridotea Tuftsii Harger (p. 340); female; dorsal view, enlarged five diameters; natural size indicated by the line at the right.

21.—The same; left maxilliped, enlarged twenty-five diameters; *e*, external lamella; *m*, basal segment; 1, 2, 3, segments of palpus.

22.—The same; pleopod of the second pair, from a male, enlarged twenty diameters; *s*, elongated stylet, articulated near the base of the inner lamella.

(All the figures were drawn from nature by O. Harger.)

Plate IV.

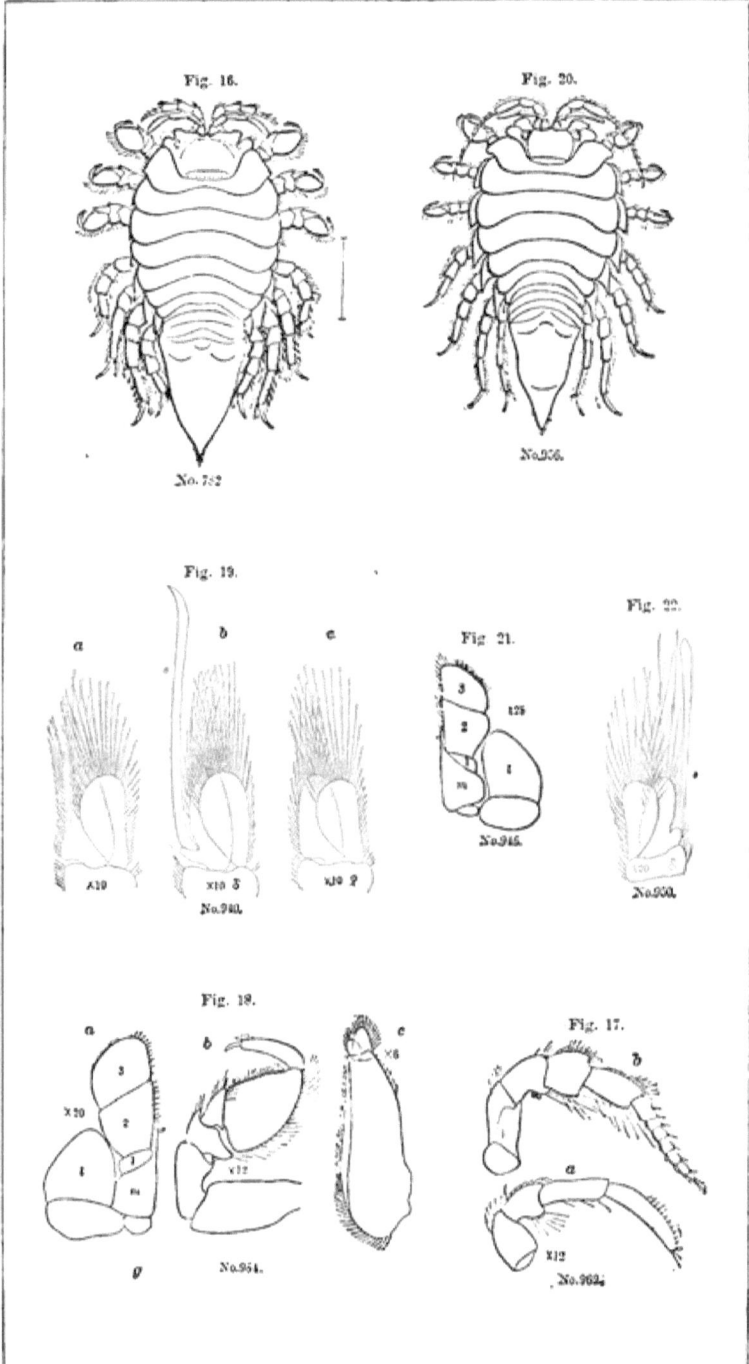

PLATE V.

FIGURE 23.—Chiridotea Tuftsii Harger (p. 340); *a*, antennula; *b*, antenna; *c*, leg of the first pair; *d*, leg of the fourth pair; all enlarged twelve diameters; *e*, left uropod, or opercular valve, inner view, enlarged ten diameters.

24.—Idotea irrorata Edwards (p. 343); dorsal view, enlarged two diameters; natural size shown by the line on the left.

25.—The same; *a*, antennula; *b*, antenna; *c*, left uropod or opercular valve, external view; all enlarged six diameters.

26.—The same; *a*, right maxilliped, enlarged twelve diameters, *l*, external lamella; *m*, basal segment; 1, 2, 3, 4, segments of palpus of maxilliped; *b*, pleopod of the second pair from a male, enlarged eight diameters, showing stylet, *s*, articulated near the base of the inner lamella.

27.—Idotea phosphorea Harger (p. 347); dorsal view, enlarged about two diameters; natural size shown by the line on the right.

28.—The same; *a*, antenna, enlarged six diameters; *b*, maxilliped, enlarged twelve diameters, showing, *l*, external lamella; *m*, basal segments; 1, 2, 3, 4, segments of the palpus of maxilliped; *c*, leg of the first pair; *d*, leg of the second pair, both enlarged six diameters; *e*, right uropod, or opercular valve, inner view, enlarged six diameters.

29.—The same; pleopod of the second pair from a male, enlarged eight diameters; *s*, stylet articulated near the base of the inner lamella; *s'*, distal end of stylet reversed and enlarged thirty diameters.

(Figure 24 was drawn by Mr. J. H. Emerton, the others by O. Harger.)

Report U. S. F. C. 1878.—Harger. Marine Isopoda. Plate V.

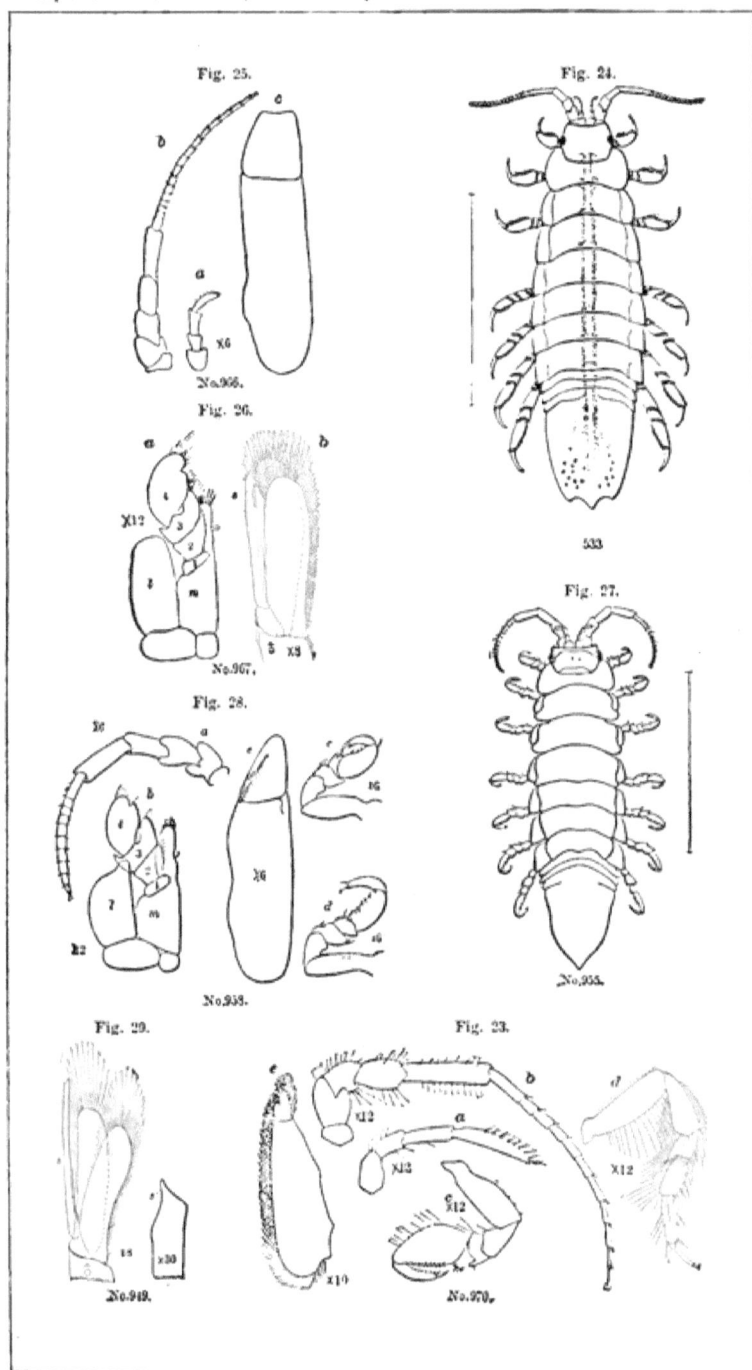

PLATE VI.

FIGURE 30.—Idotea robusta Krøyer (p. 349); dorsal view, enlarged two diameters; natural size shown by the line at the right.

    31.—The same; *a*, antenna; *b*, leg of the first pair, each enlarged six diameters; *c*, left uropod, or opercular valve, inner view, enlarged four diameters.

FIGURE 32.—The same; *a*, maxilliped, enlarged twelve diameters; *l*, external lamella; 1, 2, 3, 4, segments of palpus; *b*, maxilla of the outer or second pair; *c*, pleopod of the second pair from a male, enlarged six diameters; *s*, stylet articulated near the base of the inner lamella.

    33.—Synidotea nodulosa Harger (p. 351); dorsal view, enlarged four diameters; natural size indicated by the line at the right.

    34.—The same; *a*, antennula; *f*, flagellar segment; *b*, antenna; *c*, leg of the first pair from the right side; *d*, right uropod, or opercular valve, all enlarged ten diameters.

    35.—The same; *a*, maxilliped from the right side, showing, *l*, external lamella; *m*, basal segment; 1, 2, 3, segments of palpus, enlarged twenty diameters; *b*, maxilla of the outer or second pair; *c*, maxilla of the inner or first pair, both enlarged twenty diameters; *d*, pleopod of the second pair from a male, enlarged twelve diameters; *s*, stylet articulated near the base of the inner lamella.

    36.—Erichsonia attenuata Harger (p. 356); dorsal view, enlarged three diameters, natural size indicated by the line at the right.

(Figures 30 and 36 were drawn by Mr. J. H. Emerton, the others by O. Harger.)

Report U. S. F. C. 1878.—Harger. Marine Isopods. Plate VI.

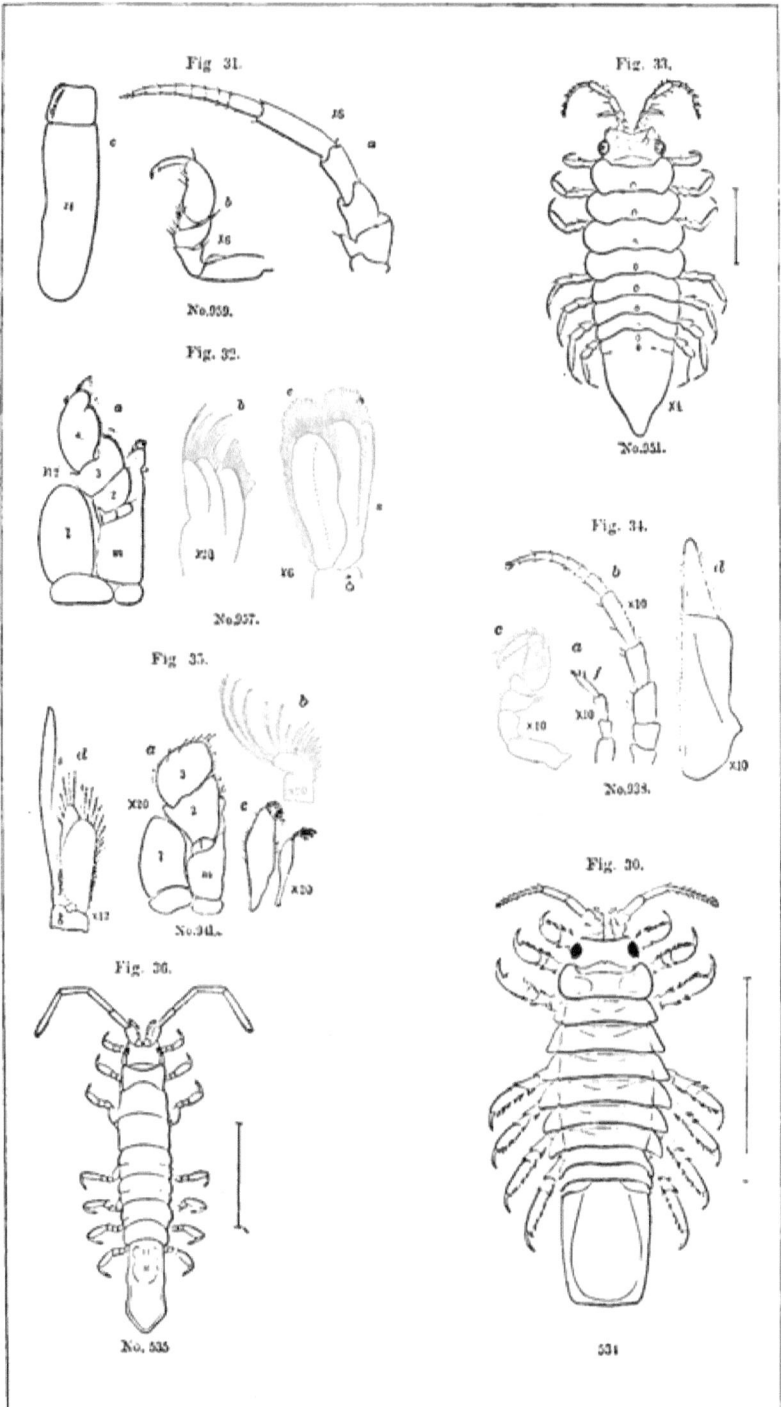

PLATE VII.

FIGURE 37.—Erichsonia attenuata Harger (p. 356); *a*, antennula; *b*, antenna, each enlarged twelve diameters; *c*, maxilliped, showing, *l*, external lamella, enlarged thirty diameters; *d*, uropod, or opercular valve, enlarged twelve diameters; *e*, pleopod of the second pair from a male, enlarged fifteen diameters; *s*, stylet, articulated near the base of the inner lamella; *s'*, distal end of stylet, enlarged fifty diameters.

38.—Erichsonia filiformis Harger (p. 355); dorsal view, enlarged five diameters, natural size indicated by the line at the right.

39.—The same; *a*, antennula; *b*, antenna; *c*, leg of the first pair; *d*, uropod, or opercular valve, each enlarged twelve diameters.

40.—The same; *a*, maxilla of outer or second pair; *b*, maxilla of inner or first pair; *c*, mandible, showing molar process, *m*, and dentigerous lamella, *d*, all enlarged thirty diameters.

41.—The same; *a*, maxilliped, showing, *l*, external lamella; *m*, basal segment, and 1, 2, 3, 4, segments of palpus, enlarged thirty diameters; *b*, pleopod of the second pair from a male, enlarged fifteen diameters; *s*, stylet, articulated near the base of the inner lamella; *s'*, distal end of stylet, enlarged fifty diameters.

42.—Epelys trilobus Smith (p. 358); dorsal view, enlarged ten diameters; natural size indicated by the line at the right.

43.—The same; *a*, maxilliped from the left side, enlarged twenty diameters; *l*, external lamella; *m*, basal segment; 1, 2, 3, segments of palpus of maxilliped; *b*, pleopod of second pair from a male, enlarged twenty diameters; *s*, stylet, articulated near the base of the inner lamella; *s'*, end of stylet, enlarged fifty diameters.

(All the figures were drawn from nature by O. Harger.)

Plate VII.

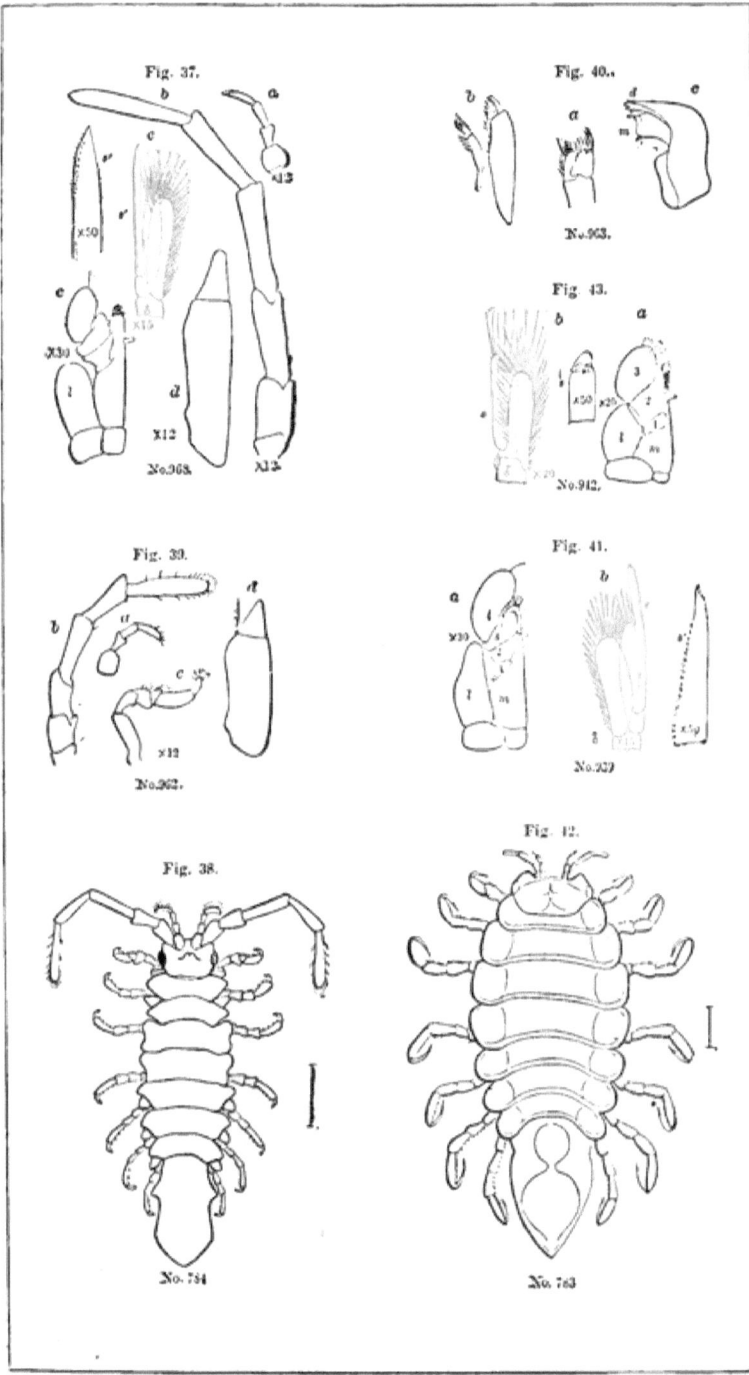

PLATE VIII.

FIGURE 44.—Epelys montosus Harger (p. 359); dorsal view, enlarged six diameters, natural size indicated by the line at the right.

45.—The same; *a*, antennula; *f*, flagellar segment; *b*, antenna; *c*, maxilliped from the left side; *l*, external lamella; *m*, basal segment; 1, 2, 3, segments of palpus; all the figures enlarged twenty diameters.

46.—The same; *a*, leg of the first pair, enlarged twenty diameters; *b*, right uropod or opercular valve, enlarged fifteen diameters.

FIGURE 47.—The same; pleopod of the second pair, from a male, enlarged twenty diameters; *s*, stylet, articulated near the base of the inner lamella; *s'*, distal end of stylet, enlarged sixty-six diameters.

48.—Astacilla granulata Harger (p. 364); female; dorsal view, enlarged four diameters, natural size indicated by the line at the right; *a*, antennula of male; *b*, fourth thoracic segment of male; *c*, inferior surface of pleon of a male, showing opercular valves; all the figures enlarged four diameters.

49.—The same; *a*, flagellum of antenna, enlarged twenty diameters; *a'*, portion of inner margin of the same, enlarged one hundred diameters; *b*, one of the first pair of legs, upper surface, enlarged twenty diameters.

50.—The same; one of the fourth pair of legs, enlarged twenty diameters.

51.—The same; inner surface of left opercular plate, or uropod, from a female, enlarged twenty diameters.

(**All** the figures were drawn from nature by O. Harger.)

Report U. S. F. C. 1878.—Harger. Marine Isopods. Plate VIII.

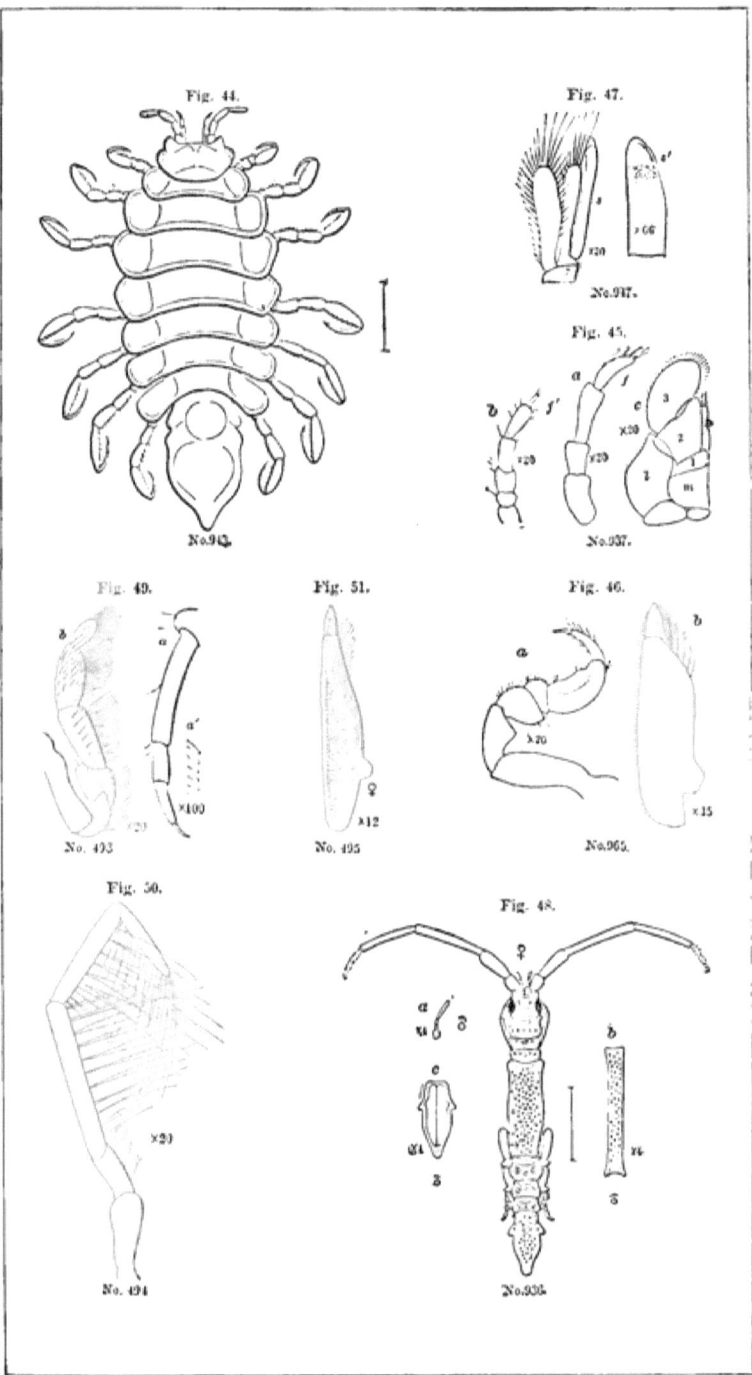

PLATE IX.

FIGURE 52.—Astacilla granulata Harger (p. 364); *a*, maxilliped; *m*, basal segment; *l*, external lamella; *b*, outer maxilla; *c*, inner maxilla; all enlarged twenty diameters.

53.—Sphæroma quadridentatum Say (p. 368); dorsal view, enlarged five diameters; natural size indicated by the line at the right.

54.—The same; *a*, antennula; *b*, antenna; *c*, pleopod of the second pair, from a male, showing stylet, *s*, articulated near the base of the inner lamella; all the figures enlarged ten diameters.

55.—Limnoria lignorum White (p. 373); dorsal view, enlarged ten diameters; natural size indicated by the line at the right.

56.—The same; *a*, antennula; *b*, antenna; *c*, maxilliped; *d*, maxilla of the outer or second pair; *e*, maxilla of the inner or first pair; *f*, mandible, all enlarged twenty-five diameters; *e'*, distal end of outer lobe of first pair of maxillæ, enlarged sixty-six diameters.

57.—The same; *a*, last segment of pleon, with attached uropods; dorsal view, enlarged ten diameters; *b*, uropod with dotted adjacent outline of last segment of pleon, enlarged thirty diameters; *c*, first pair of pleopods; *d*, pleopod of the second pair, from a male, showing stylet, *s*, articulated to the inner lamella; both figures enlarged twenty diameters.

58.—Cirolana concharum Harger, (p. 378); lateral view, enlarged about three diameters.

(Figure 53 was drawn by Mr. J. H. Emerton, 55 by Prof. S. I. Smith, 58 by Mr. J. H. Blake, and the others by O. Harger.)

Report U. S. F. C. 1878.—Harger. Marine Isopods.   Plate IX.

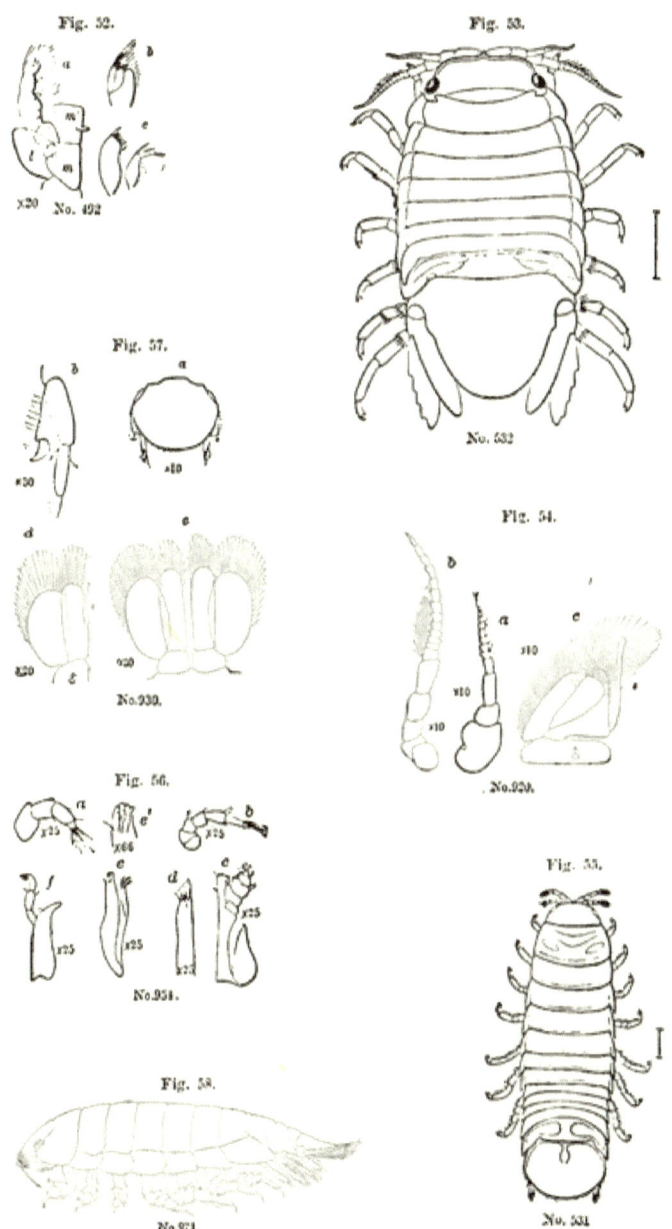

PLATE X.

FIGURE 59.—Cirolana concharum Harger (p. 378); dorsal view, enlarged about three diameters. The natural size is shown by the line at the right.
60.—The same; antennula, enlarged ten diameters.
61.—The same; *a*, antenna enlarged ten diameters; *b*, maxilla of the outer or second pair; *c*, maxilla of the inner or first pair; *d*, mandible from the right side, inner view; *p*, palpus; *m*, molar area; the last three figures enlarged five diameters.
62.—The same; *a*, maxilliped from the right side, exterior view, showing, *l*, external lamella; *m*, basal segment; 1, 2, 3, 4, 5, segments of the palpus; *b*, leg of the fourth pair; both the figures enlarged five diameters.
63.—The same; uropod from the right side; inferior view, enlarged five diameters.
64.—Ægea psora Krøyer (p. 384); *a*, dorsal and *b* ventral views of a young individual. The central line indicates the length of the specimen, natural size, which is here enlarged three diameters. Adults attain about the size of the figure.
FIGURE 65.—Nerocila munda Harger (p. 392); dorsal view of the type specimen, enlarged about four diameters. The natural size is shown by the cross on the right; *a*, uropod, enlarged six diameters.
66.—Ægathoa loliginea Harger (p. 393); type specimen; *a*, dorsal, and *b*, ventral view, enlarged four diameters. Its natural size is shown by the line between the figures.

(Figure 59 was drawn by Mr. J. H. Blake, the others by O. Harger.)

Report U. S. F. C. 1878.—Harger. Marine Isopods. Plate X.

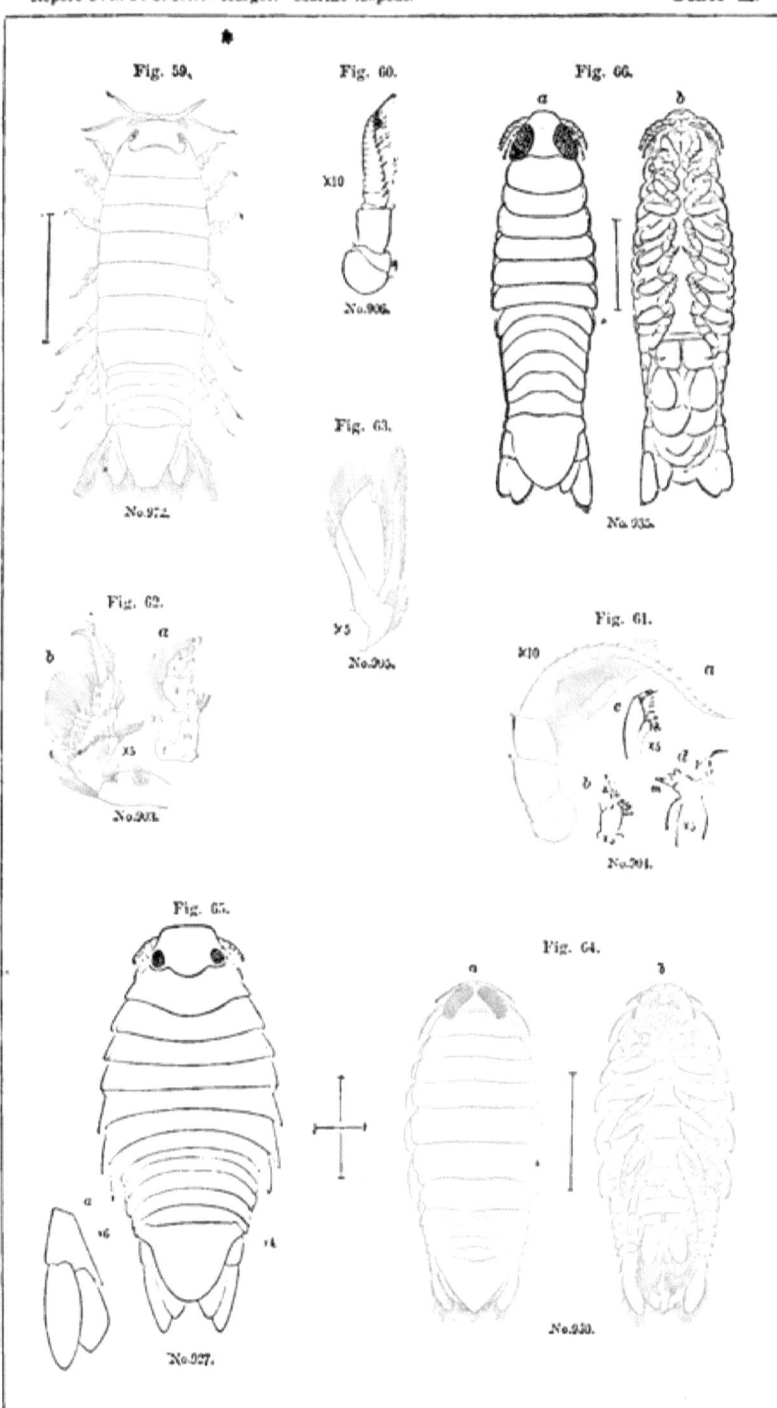

PLATE XI.

FIGURE 67.—Livoneca ovalis White (p. 305); *a*, antennula; *b*, antenna; *c*, mandibular palpus; each enlarged twenty diameters; *d*, one of the first pair of legs; *e*, one of the seventh pair of legs; *f*, uropod; each enlarged ten diameters.

68.—Anthura polita Stimpson (p. 398); dorsal view, enlarged four diameters. The natural size is shown by the line at the right; *a*, antennula; *b*, antenna, each enlarged ten diameters; *c*, leg of the first pair; *d*, leg of the third pair; *e*, right pleopod of the first pair, interior view, showing inner ramus without cilia; *f*, pleopod of the second pair from a male, showing stylet articulated to inner lamella; each of the figures *c* to *f* enlarged eight diameters; *g*, lateral view of pleon, enlarged six diameters.

69.—The same, *a*, maxilliped, enlarged twenty diameters; *b*, maxilla, enlarged twenty-five diameters; *b'*, distal end of the same, enlarged sixty diameters.

70.—Paranthura brachiata Harger (p. 402); dorsal view, enlarged about three diameters; natural size shown by the line at the right; *a*, antennula; *b*, antenna, enlarged eight diameters; *c*, right maxilliped, enlarged sixteen diameters; *d*, maxilla, enlarged sixteen diameters; *d'*, distal end of the same, enlarged fifty diameters; *e*, leg of the first pair; *f*, first pleopod from the right side, inner view, showing ciliated inner lamella; *g*, pleopod of the second pair from a male, showing stylet articulated to the inner lamella; figures *c* to *g* enlarged eight diameters.

71.—Ptilanthura tenuis Harger (p. 406); male; dorsal view, enlarged about four diameters; *a*, inferior view of the head and first thoracic segment, enlarged eight diameters; the flagellum of the antennulæ omitted; *b*, maxilliped; *c*, maxilla, each enlarged fifty diameters; *d*, first right pleopod, seen from within, showing ciliated inner lamella; *e*, second left pleopod, showing stylet *s* articulated to the inner lamella in the males.

72.—The same; one of the first pair of legs of a male, enlarged sixteen diameters.

73.—The same; female; dorsal view of the head, enlarged twenty-five diameters.

(Figure 71, excepting *b-d*, was drawn by Mr. J. H. Emerton, the others by O. Harger.)

Report U. S. F. C. 1878.—Harger. Marine Isopoda. Plate XI.

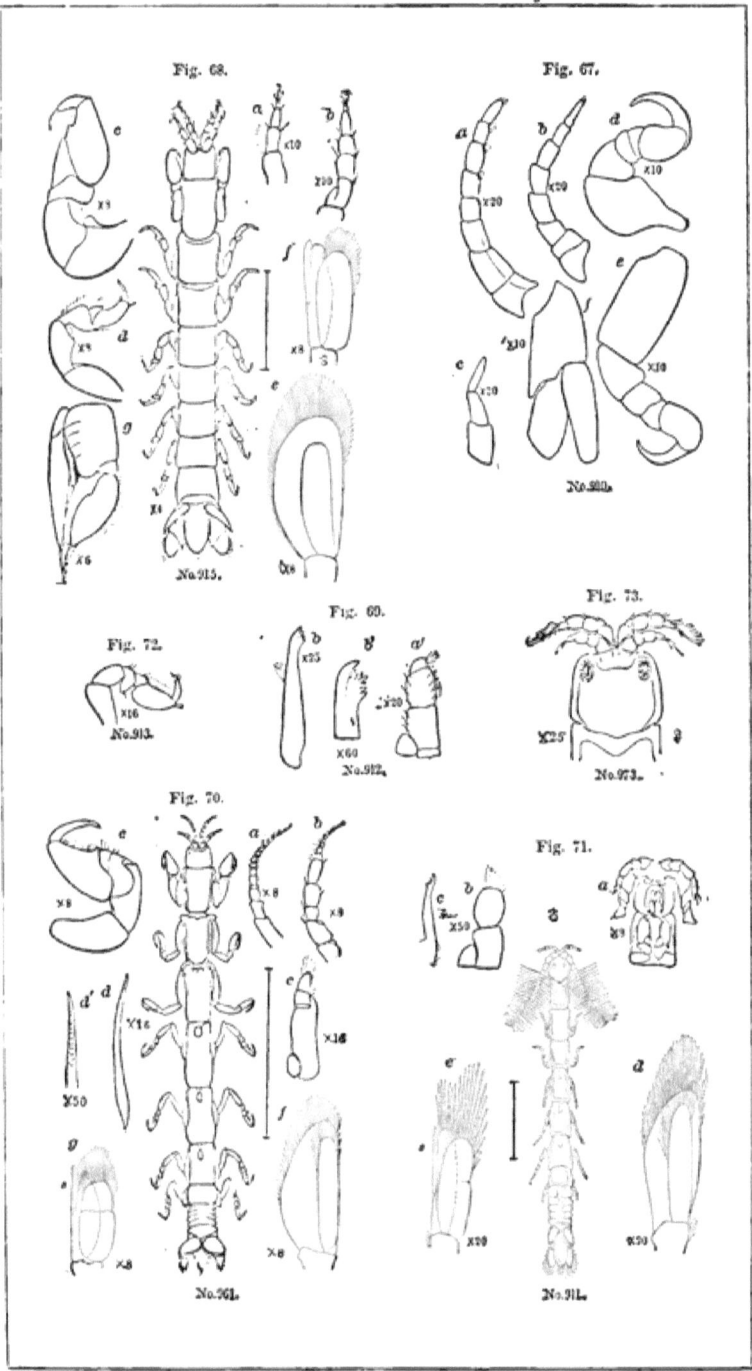

PLATE XII.

FIGURE 74.—Ptilanthura tenuis Harger (p. 406); *a*, antennula; *b*, antenna; each enlarged twenty diameters, from a male.

75.—Gnathia cerina Harger (p. 410); male; dorsal view, enlarged ten diameters.

76.—The same; *a*, antennula; *b*, antenna, each enlarged thirty-eight diameters; *c*, mandibles (*l*, left, *r*, right), enlarged thirty-eight diameters; *d*, first leg or first gnathopod from the right side, enlarged twenty-five diameters; all the figures from the male sex.

77.—The same (p. 411); female; dorsal view, enlarged ten diameters.

78.—The same; *a*, one of the first pair of legs or first gnathopod of a female, enlarged thirty-eight diameters; *b*, one of the first pair of legs in a young, parasitic individual, enlarged sixty diameters; *c*, pleon, with the last and part of the penultimate thoracic segments of a female, dorsal view, enlarged twenty diameters; *d*, pleopod of a young, parisitic individual, enlarged sixty diameters; *e*, pleopod of an adult male, enlarged sixty diameters.

79.—The same; young male; dorsal view, enlarged twenty diameters.

80.—Leptochelia algicola Harger (p. 421); male; lateral view, enlarged twenty diameters; natural size indicated by the line above.

(All the figures were drawn from nature by O. Harger.)

Report U. S. F. C. 1878.—Harger. Marine Isopods. Plate XII.

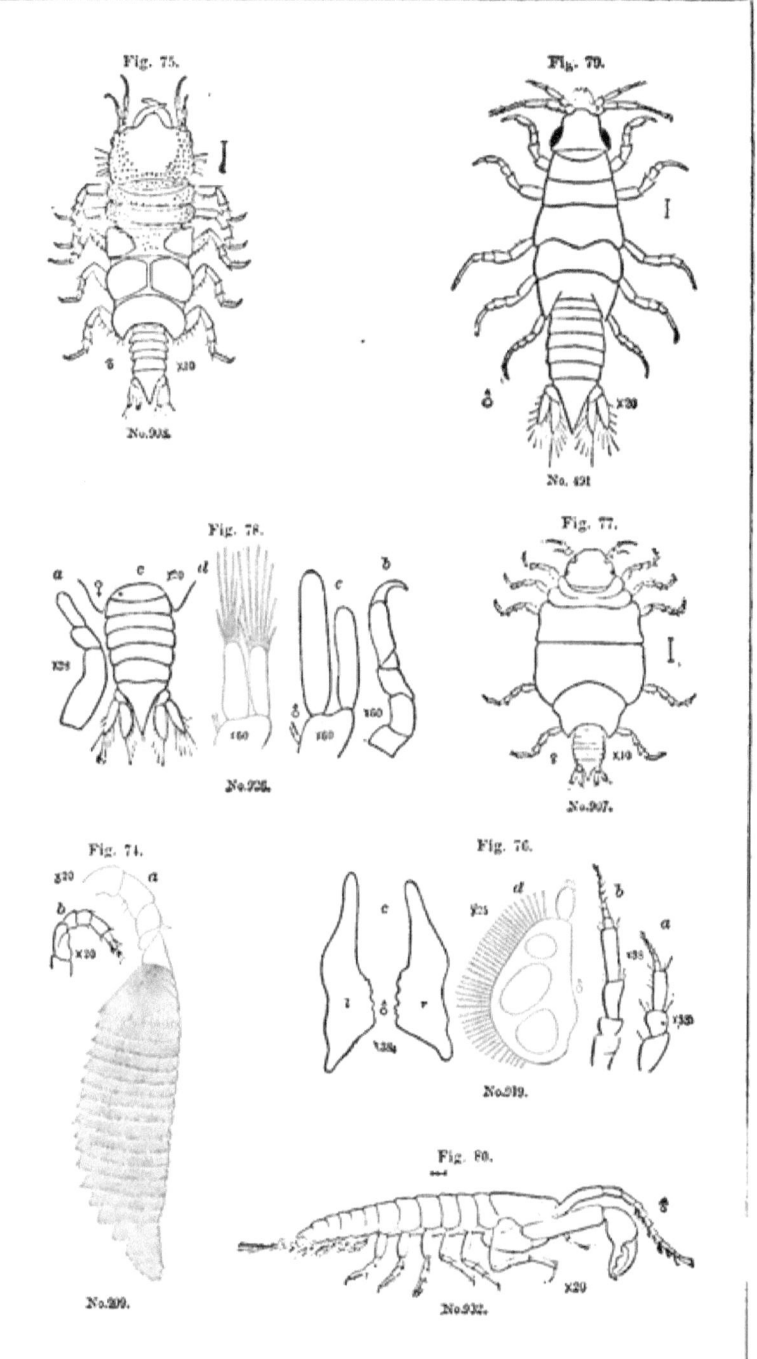

PLATE XIII.

FIGURE 81.—Tanais vittatus Lilljeborg (p. 418); dorsal view, enlarged eight diameters. The transverse bands of hairs on the pleon are not sufficiently distinct.

82.—The same; one of the first pair of pleopods, enlarged thirty diameters.

83.—Leptochelia algicola Harger (p. 421); female; dorsal view, enlarged twenty diameters; natural size indicated by the line at the right.

84.—The same; a, antennula; b, one of the first pair of legs; both from a female specimen and enlarged twenty-five diameters.

85.—The same; hand, or propodus and dactylus of the first pair of legs, enlarged forty-eight diameters, showing the comb of setæ on the propodus.

86.—The same; uropods of a male, enlarged seventy diameters; b, basal segment; i, inner six-jointed ramus; o, outer ramus.

87.—Leptochelia limicola Harger (p. 424); female; dorsal view, enlarged twenty diameters; natural size shown by the line at the right.

88.—The same; a, antennula; b, antenna; c, leg of the first pair; d, leg of the second pair; all from the female sex and enlarged twenty-five diameters.

89.—Leptochelia rapax Harger (p. 424); male; dorsal view, enlarged about twelve diameters.

90.—The same; hand, or propodus and dactylus of male, enlarged sixteen diameters.

91.—Leptochelia coeca Harger (p. 427); type specimen, female; a, antennula; b, leg of the first pair; c, uropod; each enlarged fifty diameters.

(All the figures were drawn from nature by O. Harger.)

Report U. S. F. C. 1878.—Harger. Marine Isopods. Plate XIII.

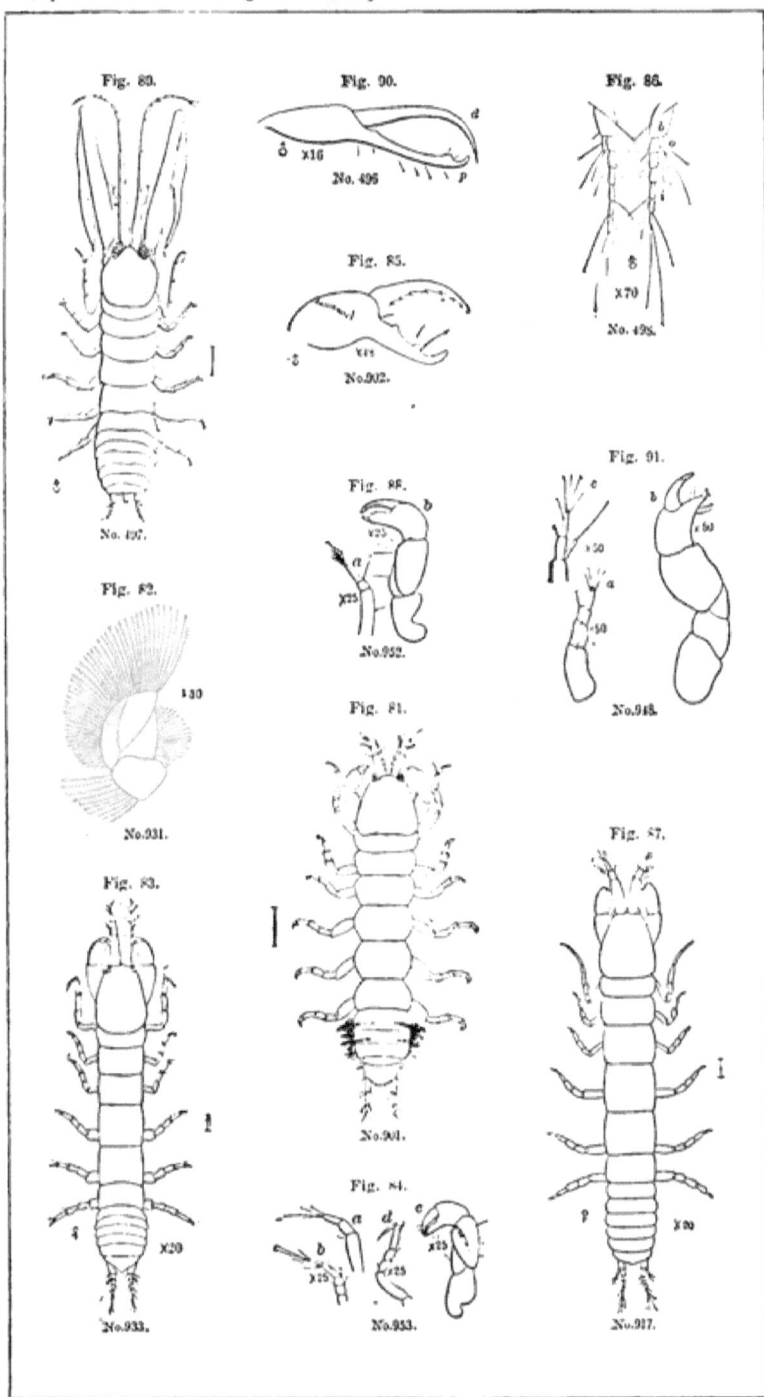

ALPHABETICAL INDEX TO THE REPORT ON THE MARINE ISOPODA OF NEW ENGLAND AND ADJACENT WATERS.

[In the following index the first reference for the names of the families, genera, and species here described is to the page on which such description is made. The list of authorities, being alphabetically arranged, is not indexed.]

Abdomen, 298.
Actæcia, 309.
Actoniscus, 309, 305.
    ellipticus, 309, 428, 433.
Æga, 383, 377, 378, 387, 431.
    concharum, 378.
    emarginata, 384.
    entaillée, 384.
    polita, 381.
    psora, 384, 429, 430, 434.
Ægathoa, 393, 391.
    loliginea, 393, 428, 434.
Ægidæ, 382, 300, 303, 377, 430, 431.
Æro-spirantia, 305.
Alitropus, 391.
Anceus, 409, 410.
    americanus, 410, 411.
    elongatus, 412.
Ancoral legs, 300.
Andrews, A., on Limnoria, 375.
Anilocra, 391.
    mediterranea, 430.
Anisocheirus, 416.
Antennæ, 298.
Antennulæ, 298.
Anthura, 398, 301, 397, 431.
    brunnea, 398, 401.
    carinata, 401.
    gracilis, 398, 401.
    polita, 398, 400, 429, 434.
Anthuridæ, 396, 301, 303, 361, 431.
Apseudes, 304, 414, 416, 431.
Arcturidæ, 361, 303, 397, 430, 431.
Arcturus, 361, 363.
    Baffini, 362.
Armadillidæ, 314.
Armadillo, 305.
Armida bimarginata, 343.
Artystone, 390.
Asellidæ, 312, 303, 314, 371, 372, 430, 431.
Asellodes, 319.
    alta, 319, 321.
Asellotes homopodes, 371.
Asellus, 301, 313, 415.
    communis, 314.
    Grönlandicus, 315, 319.
Astacilla, 361, 297, 301, 363, 431.
    Americana, 364.
    granulata, 364, 362, 429, 434.
    longicornis, 362, 363, 366.
Basis, 300.
Bate, C. Spence, on the incubatory pouch, 302.

Bate, C. Spence, on terminology of crustacea, 300.
Bate and Westwood, on Æro-spirantia, 305. [302.
    Anceus, 409.
    Anthura, 398.
    Anthura gracilis, 401.
    British Isopoda, 429.
    Idotea tricuspidata, 345.
    Jæra albifrons, 318.
    Limnoria, 372.
    Paranthura, 402.
    Tanais, 416.
    Tanais Edwardsii, 422.
Bathynomus giganteus, 383.
Bopyridæ, 311, 377, 429, 431.
Bopyrus, 312, 431.
    abdominalis, 312.
    Hippolytes, 311.
    Mysidum, 312.
    species, 312, 428, 433.
Brevoortia menhaden, 391.
Bullar, J. F., hermaphroditism in Cymothoidæ, 391.
Cancer maxillaris, 410.
Carpus, 300.
Cepon distortus, 311, 428, 433.
Ceratacanthus, 393.
Chætilia, 336.
Chela, 300.
Chelura terebrans, 371, 376, 419, 423.
Chiridotea, 337, 300, 335, 336.
    cœca, 338, 335, 340, 429, 433.
    entomon, 337.
    Tuftsii, 340, 429, 433.
Cirolana, 378, 376, 383, 431.
    concharum, 378, 298, 428, 434.
    polita, 381, 429, 434.
    truncata, 430.
Cirolanidæ, 376, 303, 382, 430, 431.
Cleantis, 336.
Cœcidotea, 314.
Coldstream, J., on Limnoria, 372.
Conilera, 376, 378.
    concharum, 378.
    polita, 381.
Cordiner, C., on Astacilla, 363.
Coxa, 300.
Crossurus, 416.
    vittatus, 416, 418.
Cuma, 415.
Cymothoa, 383, 391.
    œstrum, 377.
    ovalis, 395.
    prægustator, 391.

459

Cymothoæ, 377.
Cymothoidæ, 290, 300, 303, 371, 382, 430, 431.
Dactylus, 300.
Dajus Mysidis, 312, 429, 433.
Dana, J. D., on Asellidæ, 314.
Desmarest, A. G., on Idotea tricuspidata, 345.
Digital process, 300.
Dohrn, A., on Anceus, 409.
    the incubatory pouch, 301, 303.
    Tanaidæ, 415.
Edriophthalma, 297.
Edwards, H., on Idotea tricuspidata, 345.
    Limnoria, 371.
    Tanais, 416.
Epelys, 357, 301, 337.
    montosus, 359, 429, 434.
    var. hirsutus, 360.
    trilobus, 358, 429, 434.
Epimera, 300.
    abdominal, 302.
Erichsonia, 354, 337, 361.
    attenuata, 356, 335, 428, 434.
    filiformis, 355, 428, 434.
Eurycope, 332, 329.
    cornuta, 333.
    robusta, 332, 429, 433.
Flagellum, 298.
Fleming, J., on Astacilla, 363.
Gadus, 386.
Gammarus Dulongii, 416.
Gegenbaur, C., on Tanaida, 415.
Gelasimus pugilator, 311.
Geographical distribution, 428.
Gnathia, 410, 297, 302, 357, 431.
    cerina, 410, 429, 435.
    termitoides, 410.
Gnathiidæ, 408, 300, 301, 303, 431.
Gnathium, 410.
Gnathopods, 300.
Goodsir, H. D. S., on Astacilla, 363.
Gribble, 375.
Gyge, 431.
Gyge Hippolytes, 311, 429, 433.
Helleria, 305.
Henopomus tricornis, 322.
Hesse, E., on Anceus and Praniza, 409.
Hippoglossus, 382, 386.
Hippolyte Fabricii, 311.
    polaris, 311.
    pusiola, 311, 312.
    securifrons, 311, 312.
    spinus, 311, 312.
Huxley, T. H., distinction of cephalic and thoracic
Idotea marina, 214.   [segments, 302.
Idotea, 311, 337, 431.
    acuminata, 344.
    Basteri, 343.
    bicuspida, 352.
    cæca, 338.
    entomon, 345.
    filiformis, 355.
    irrorata, 343, 342, 348, 429, 430, 433.
    marina, 344.
    marmorata, 352.
    metallica, 349, 350.
    montosa, 359.

Idotea nodulosa, 352.
    pelagica, 343, 345, 346.
    phosphorea, 347, 342, 346, 429, 433.
    pulchra, 352.
    robusta, 349, 342, 429, 433.
    tricuspidata, 343, 345, 346
    tricuspis, 344.
    tridentata, 344, 345.
    Tuftsii, 346.
    variegata, 343.
Idoteidæ, 335, 301, 303, 361, 397, 430, 431.
Idothea balthica, 344, 346.
    nodulosa, 351.
    pelagica, 344.
    robusta, 349.
Ilyarachna, 334, 329, 335, 429, 433.
Incubatory pouch, 301.
Ischium, 300.
Isopod, length of, 302.
Isopoda, 297.
    aberrantia, 303.
Jæra, 314, 301, 313, 430, 431.
    albifrons, 315, 318, 429, 430, 433.
    Baltica, 315, 318.
    copiosa, 315.
    Kröyeri, 315, 318.
    maculata, 315, 318.
    marina, 315, 318.
    nivalis, 315, 318.
    triloba, 358.
Janira, 319, 313, 430, 431.
    alta, 321, 299, 429, 433.
    laciniata, 324.
    maculosa, 319, 322.
    spinosa, 323, 429, 433.
Johnston, G., on Astacilla, 362.
Kinahan, J. R., on Aetæcia, 309.
Kröyer, H., on Anthura carinata, 401.
    Munna, 325.
    Tanais Edwardsii, 422.
Labium, 300.
Labrum, 300.
Lamella, external of maxillipeds, 299.
Latreille, P. A., on Idotea tricuspidata, 345.
Leachia, 361, 363.
    granulata, 364, 366.
Leach, W. E., on Gnathia, 410.
Leacia, 361, 363.
Legs, 300.
Leidya distorta, 311.
Leidy, J., on Bopyrus species, 312.
    C pon distortus, 311.
Leptochelia, 420, 301, 414, 415, 431.
    algicola, 421, 429, 435.
    cæca, 427, 429, 435.
    Edwardsii, 416, 421, 422, 423.
    filum, 426, 429, 435.
    limicola, 424, 429, 435.
    minuta, 416.
    rapax, 424, 423, 429, 435.
Leptophryxus Mysidis, 312.
Ligia, 305, 310, 311, 415.
Lilljeborg, W., on Jæra albifrons, 318.
Limnoria, 373, 313, 371, 419, 430, 431.
    lignorum, 373, 429, 430, 434.
    terebrans, 373.

ALPHABETICAL INDEX.

Limnoria uncinata, 374, 376.
    xylophaga, 371.
Limnoriadæ, 371.
Limnoriidæ, 371, 303, 431.
Lironeca, 395.
Livoneca, 394, 383.
    ovalis, 395, 300, 428, 434.
Lockington, W. N., on color of Idotea pulchra, 353.
Loligo Pealii, 394.
Macdonald, J. D., on Tanais vittatus, 417, 419.
Mancasellus, 313.
Mandibles, 299.
Maxillæ, 299.
Maxillipeds, 299.
Mayer, P., Hermaphroditism in Cymothoidæ, 391.
Meinert, F., on Idotea tricuspidata, 346.
Merus, 300.
Mesostenus, 334.
Metzger, A., on Jæra albifrons, 318.
Micropogon undulatus, 396.
Möbius, K., on Jæra marina, 318.
Molar process, 299.
Montagu, G., on Oniscus gracilis, 401.
Müller, F., on Bopyridæ, 303, 311.
    Leptochelia and Paratanais, 420.
    olfactory setæ, 298.
    Tanaidæ, 303, 415.
    Tanais Edwardsii, 423.
Mullet, 394.
Munna, 325, 313, 430, 431.
    Boeckii, 325, 328.
    Fabricii, 325, 429, 433.
Munnopsidæ, 328, 303, 314, 430.
Munnopsis, 329.
    typica, 330, 334, 429, 433.
Nerocila, 391.
    munda, 392, 428, 434.
Norman, A. M., British Cymothoidæ, 430.
Ocelli, 298.
Olfactory setæ, 298.
Oniscidæ, 305, 303, 311, 314.
Oniscoidea, 314.
Oniscus albifrons, 315.
    Balthicus, 344.
    cœruleatus, 410.
    entomon, 344.
    gracilis, 401.
    marinus, 318.
    praegustator, 391.
    psora, 384, 386.
Operculum, 302, 336.
Ourozeuktes, 377.
Palæmonetes vulgaris, 312.
Palpus, 299.
Pandalus borealis, 312.
    Montagui, 312.
Paranthura, 402, 398, 431.
    arctica, 405.
    brachiata, 402, 429, 435.
    costana, 405.
    norvegica, 404.
    tenuis, 406, 407.
Paratanais, 416, 420, 431.
    algicola, 419, 421.
    cœca, 427.
    forcipatus, 423.

Paratanais limicola, 424, 427.
Peduncle, 298.
Pereion, 298, 300.
Pereiopods, 300.
Pleon, 298, 301.
Pleopods, 301.
Philoscia, 305.
    vittata, 306, 429, 433.
Phryxus, 431.
Phryxus abdominalis, 312, 429, 433.
    Hippolytes, 312.
Pill-bug, 298, 305.
Platyarthrus, 308.
Pomatomus saltatrix, 396.
Porcellio, 305.
Praniza, 409, 410.
    cerina, 410, 412.
    cœruleata, 410.
    Reinhardi, 413.
Propodus, 300.
Ptilanthura, 405, 398.
    oculata, 408.
    tenuis, 406, 429, 435.
Raia, 386.
Rathke, H., on Crossurus, 416, 417.
Respiration, 302, 303.
Rostrum, 302.
Salve-bug, 384.
Sars, G. O., on Eurycope, 332.
    Ilyarachna, 334.
    Munnopsidæ, 329.
    Munnopsis, 331.
Sars, M., on Idotea tricuspidata, 346.
    Jæra albifrons, 318.
    Munnopsidæ, 329.
    Munnopsis, 331.
Schiödte, J. C., on Anthura, 397.
    Artystone, 390.
    Cymothoa, 377.
Scyphacella, 307.
    arenicola, 307, 428, 433.
Scyphax, 307.
    ornatus, 309.
Serolids, 304.
Smith, S. I., list of Bopyridæ, 311.
    Limnoria xylophaga, 371.
    Scyphacella, 307.
Sow-bug, 298, 299, 305.
Sphæroma, 368, 301, 367, 372, 430, 431.
    quadridentatum, 368, 429, 434.
    serratum, 430.
Sphæromidæ, 367, 303, 431.
Stebbing, T. R. R., on Astacilla, 362.
    British Arcturidæ, 430.
    Dynamene rubra and viridis, 430.
    Tanais vittatus, 417, 419.
Stenosoma filiformis, 355.
    irrorata, 343.
Stenotomus argyrops, 396.
Stimpson, W., on Anceus americanus, 410, 413.
    Asellodes, 319.
    Cirolana concharum, 381.
    Cirolana polita, 381.
    Praniza cerina, 412, 413.
    Tanais filum, 426.

Synidotea, 350, 337.
    bicuspida, 352, 429, 433.
    nodulosa, 351, 299, 347, 429, 433.
Syscenus, 387, 383, 391.
    infelix, 387, 429, 434.
Tanaida, 415.
Tanaidæ, 413, 298, 300, 302, 303, 304, 431.
Tanais, 416, 297, 301, 414, 415, 431.
    Cavolinii, 416, 419.
    Dulongii, 416.
    Edwardsii, 416, 421, 423.
    filum, 420, 421, 423, 426.
    hirticaudatus, 418.
    islandicus, 428.
    Savignyi, 423.

Tanais, tomentosus 418, 419.
    vittatus, 418, 417, 420, 428, 429, 431, 435.
Telson, 301.
Templeton, R., on Zeuxo, 416.
Tetradecapoda, 297.
Thorax, 298, 300.
Tylus, 305.
Uropods, 301.
Verrill, A. E., on Chiridotea Tuftsii, 341.
Venus mercenaria, 359.
Westwood, J. O., on Anisocheirus, 416.
White, A., on Limnoriadæ, 371.
Willemoes-Suhm, R. v., on Tanais, 418.
Wood-lice, 305.
Zeuxo Westwoodiana, 416.

[Report of the United States Commissioner of Fish and Fisheries. Part vi. For 1878. Opposite page 462.]

ERRATA.

Page 313,	line	5,	for	79	read	371.
" 329,	"	22,	"	38	"	352.
" "	"	23,	"	40	"	354.
" 383,	"	20,	"	89	"	385.
" "	"	" "	"	93	"	387.
" 398,	"	9,	"	104	"	398.
" "	"	11,	"	108	"	402.
" "	"	12,	"	111	"	405.
" 416,	"	35,	"	122	"	416.
" "	"	36,	"	126	"	420.
" 434,	"	55,	"	139	"	433.
" "	"	" "	"	141	"	435.
" 435, last line,	"	157	"	451.

In partial explanation of the above list of incorrect references, the author has to say that he had no opportunity of seeing a correctly paged proof of the article during the year that it was in press at the Government Printing Office.

O. H.

Transpose explanations of figs 84 & 88 Pl. XIII.

Descriptions of new Genera and Species of Isopoda, from New England and Adjacent Regions; by OSCAR HARGER. *Brief Contributions to Zoology from the Museum of Yale College.*

THE genera and species described in the present paper are, except the first, marine and were, mostly, collected by the United States Fish Commission, along the New England coast. More complete descriptions with figures of all the new, and most of the old species, are nearly ready for publication in the Report of the Commissioner. As it seems desirable, however, to give a wider publication to the genera and species believed to be new, the following diagnoses are here inserted.

Actoniscus, gen. nov.*

Eyes small. Antennæ geniculate at the third and fifth segments; flagellum four-jointed. Terminal segment of maxillipeds lamelliform. Legs all alike. Pleon of six distinct segments. Basal segments of uropoda dilated and simulating the coxæ of the preceding segments; rami both styliform.

This genus belongs to the *Oniscidæ* and is near *Actæcia* Dana, MSS. (U. S. Expl. Exped., Crust., part II, p. 736, pl. 48, fig. 6), regarded as the young of *Scyphax*, but considered by Kinahan as the representative of a distinct family of the Oniscoidea.

A. ellipticus, n. sp. Body oval. Head with a prominent angular median lobe, and broadly rounded, divergent lateral lobes. Eyes oval, longitudinal, prominent, black. Antennulæ rudimentary. Antennæ nine-jointed: first segment short; second strongly clavate; third smaller, clavate; fourth flattened-cylindrical; fifth longest, slender, bent at the base; flagellum shorter than the fifth segment, composed of four subequal segments, tipped with setæ. Terminal segment of maxillipeds elongate triangular, ciliated and slightly lobed near the tip. First thoracic segment excavated in front for the head, shorter above than the following segments except the last, which is shortest. Legs small, scarcely spiny. Pleon continuing the regular oval outline of the thorax, apparently with four pairs of lamellar coxæ, the last pair are, however, the enlarged basal segments of the uropoda and are notched on their inner margins for the short outer rami, while the more slender inner rami are borne lower down on the under surface. The rami scarcely project beyond the general outline.

This species has been collected by Professor A. E. Verrill, at Savin Rock, near New Haven, and also at Stony Creek, in company with *Philoscia vittata* Say.

* From ἀκτή, the beach, and *Oniscus.*

Chiridotea,* gen. nov.

First three pairs of legs terminated by prehensile hands, in each of which the carpus is short and triangular, the propodus is robust and the dactylus capable of complete flexion on the propodus. Antennæ with an articulated flagellum. Head dilated laterally. Operculum vaulted, with two apical plates.

This genus is founded on *Ch. cœca* (*Idotea cœca* Say), which occurs on this coast from Florida to Halifax, Nova Scotia. It includes *Ch. Tuftsii* (*Idotea Tuftsii* Stimpson), of the New England coast from Long Island Sound to the Bay of Fundy, and, as constituted above, would also include *Ch. entomon* (*Idotea entomon* Bosc.), from the Baltic and other European localities, and *Ch. Sabini* (*Idotea Sabini* Kröyer), from the Arctic. The above mentioned species ought certainly to be separated from *Idotea tricuspidata* Desm., which may properly be regarded as the type of the genus *Idotea* Fabr.

Synidotea,† gen. nov.

Antennæ with an articulated flagellum. Epimeral sutures not evident above. Pleon apparently composed of two segments, united above but separated at the sides by short incisions. Operculum with a single apical plate. Palpus of maxillipeds three-jointed.

This genus is founded on *S. nodulosa* (*Idotea nodulosa* Kröyer), who appears to have been misled, in his unnatural description of the epimera, by the marginal thickening of the segments. He describes the epimera as evident even on the first segment.

Astacilla Americana, sp. nov.

Body nearly uniform in size throughout in the female, with the fourth thoracic segment narrow in the male, tuberculated. Head united with the first thoracic segment, and, together with it, twice the length of the next two segments: excavated in front, with the sides extending beyond the middle of first antennular segment, tuberculated above, crossed between and behind the eyes by two transverse grooves, while a third groove indicates the suture of the first thoracic segment. Eyes lateral, round-ovate, broadest in front. Antennulæ four-jointed, slightly surpassing the second segment of the antennæ in the female, nearly attaining the middle of the third in the male; basal segment swollen, nearly as long as the next two which are much more slender, last or flagellar segment shorter than the peduncle in the female, longer than the peduncle in the male. Antennæ about three-fourths as long as the body, fourth segment longest, then the fifth and third; first two segments short; flagellum three-jointed, short. First thoracic

---

* From χείρ, a hand, and Idotea.   † From σύν, with or together, and Idotea.

segment embracing the head, separated from it by sutures at the sides, but united in the dorsal region. Fourth segment longer than the other six segments together in the female, still more elongated in the male, in which sex it is longer than the three following segments with the pleon, while in the female it is only four-fifths as long; irregularly but not coarsely tuberculated, especially in the dorsal region. Last three segments with their epimeral regions produced into salient angular tubercles. Pleon elongate-ovate, tuberculated, a little longer than the last three thoracic segments, with three transverse grooves in the proximal region, the second continued at the sides, but showing no distinct suture. Immediately behind this is a prominent tooth on each side, directed outward and backward. Tip of pleon not spiniform but only slightly attenuated and obtuse. Opercular plates more than nine-tenths as long as the inferior surface of the pleon.

Length of female 10mm., male 11mm.; diameter of fourth thoracic segment, female 1·2mm., male 0·52mm. Of the two adult specimens obtained, fortunately a pair, the male, though much the more slender, actually exceeds the female in length. This relation of size in the sexes is unusual in the genus, the females being generally considerably larger than the males, but more specimens are necessary to prove the constancy of this proportion.

The specimens of this species were found adhering to *Primnoa*, from St. George's Bank.

Astacilla Fleming, is synonymous with *Leacia* (*Leachia*) Johnston, which is preoccupied.

Eurycope robusta, sp. nov.

Body oval, smooth and polished, breadth nearly equal to half the length. Head longer than the first thoracic segment, produced medially into a short rostrum about half as long as the basal antennular segments. Antennulae attaining the middle of the fourth antennal segment; basal segment subquadrate, bearing the second, much smaller, segment beyond the middle of its superior surface; third segment slender; flagellum about twice as long as the peduncle, multiarticulate. Antennae thrice the length of the body at least in the female; first three segments short; fourth and fifth slender, subequal and together as long as the body in the female; flagellum long, slender and multiarticulate. External lamella of maxillipeds subrhombic, with the inner angle much rounded, the outer prominent but not acute. First four thoracic segments short; fourth widest, fifth and sixth suddenly twice as long; seventh much the longest of all. First pair of legs shorter than the body; carpus exceeding the propodus; second pair longer than the body; third and fourth increasing slightly in length; carpus and propodus subequal in all, armed, in the second pair only, with spines. Swimming legs (last three pairs) robust, carpus subcircular, dactylus usually about half as long as the propodus. Pleon broader than long. Uropoda short, rami cylindrical, spiny at the tip; the outer more slender but not shorter than the inner. Length of body 4·5mm. Carpus of first pair 1mm.; propodus 0·6mm.; of second pair, carpus 1·5mm., propodus 1·6mm.; of fourth pair, carpus 1·5mm., propodus 1·7mm. Color, in alcohol, pale honey yellow.

This species was dredged in 220 fathoms, in the Gulf of St. Lawrence, by Mr. J. F. Whiteaves.

Ægathoa loliginea, sp. nov.

Body elongate oval, not suddenly narrower at the base of the pleon, which is slightly dilated at the last segment. Head subequally, but not deeply, lobed behind, the lateral lobes being formed by the large, semi-hexagonal, coarsely reticulated eyes, which cover half the upper surface of the head. Antennulæ as long as the head, eight-jointed, separated at their bases, tapering to the tip; antennæ more slender, ten-jointed, surpassing the antennulæ by the last two segments, like the antennulæ without evident division into peduncular and flagellar segments. First thoracic segment shorter than the head and but little broader, not embracing it at the sides, longer than the following segments, which increase in width to the fifth; seventh shortest. Epimera short and obtuse, not surpassing the rounded posterior angles of the segments. Legs nearly alike throughout, first pair a little more robust, last pair slightly the longest, all with strongly curved dactyli. Pleon longer than the thorax, tapering to the fifth segment. First pair of pleopoda with the basal segment large, nearly square; last pair, or uropoda, surpassing the telson; basal segment triangular with the inner angle acute but scarcely produced; rami flat, the outer with slightly divergent sides, obliquely rounded at the end; the inner broader, triangular, with the outer side longest; cilia very short almost rudimentary. Length 13mm., breadth 3·6mm. Color in alcohol yellowish with minute black specks, most abundant on the pleon. Eyes black.

The only specimen in the collection was obtained by Mr. S. F. Clark, at Savin Rock, near New Haven, from the mouth of a squid (*Loligo Pealii*), whence the specific name.

Ptilanthura, gen. nov.*

Antennulæ with the flagellum remarkably developed, multiarticulate, second and following segments provided with an incomplete, dense whorl of fine slender hairs. This whorl is interrupted in each segment upon its internal or anterior side,

* From πτίλον a plume, and Anthura.

which, however, in the ordinary reflexed position becomes the external side. Eyes distinct, visible both above and below. Pleon imperfectly segmented, elongate. Maxillipeds two-jointed.

P. tenuis, sp. nov. Body smooth, slender, flattened above, broadest at the base of the pleon. Head broader but shorter than the first thoracic segment, narrowed to a point in front and less acutely behind. Eyes prominent, black, within the margin of the head. Antennulæ, when reflexed, attaining the third thoracic segment; first segment large but not longer than the second; third shorter than the second, followed by a short first flagellar segment, second and following segments about twenty in number, obconic, fitting into each other, flattened and naked on one side, which is the outer and somewhat inferior side in the reflexed organ, densely elongate-ciliate distally, except on the flattened side; cilia attaining about the fifth following segment. Antennæ hardly surpassing the peduncle of the antennulæ, eight-jointed. Maxillipeds with a quadrate basal segment, emarginate externally for the subtriangular external lamella, and bearing a single scarcely smaller terminal segment, truncate and ciliate at the tip. Thoracic segments slender, margined, the seventh but little over half as long as the others. First pair of legs moderately enlarged, segments well separated, dactylus strong, shorter than the inner margin of the propodus; remaining pairs of legs slender. Pleon about as long as the last three thoracic segments, first five segments consolidated along the median line, each rising into a low broad tubercle on each side of the median line; last segment as long as the preceding five; telson elongate-ovate obtusely pointed. Uropoda equaling the telson. Length 11mm., breadth 0·9mm., color in life brownish and somewhat mottled above, lighter below.

This species has been found on the New England coast from Noank Harbor, Conn., to Casco Bay, Maine.

Paratanais algicola, sp. nov.

Tanais filum Harger, Rep. U. S. Com. Fish and Fisheries, part 1, p. 573, 1874, *non* Stimpson.

Eyes conspicuous, black, plainly articulated, larger in the males. Antennulæ in the females three-jointed, tapering, setose at the tip, first segment as long as the last two which are subequal; elongated and eleven-jointed in the male, the first segment long, curved upward near the base, last eight segments with olfactory setæ. Antennæ short, five-jointed, deflected, fourth segment longest. First pair of legs robust, hand short and stout in the female, digital process scarcely toothed, bearing three setæ near its inner margin; hand in males strongly chelate, digital process elongated, curved, two-toothed; dactylus curved, slender, with about seven setiform spines on its inner margin; carpus in the males long and stout.

Second pair of legs elongated, basis flattened and curved, dactylus slender but shorter than the propodus. Bases of last three pairs of legs swollen. Uropoda bearing setæ at the tips of the segments, biramous; outer ramus short, scarcely if at all surpassing the basal segment of the inner ramus which is six-jointed and tapering. Length 2·2mm., breadth 0.33mm. Color nearly white.

This species is rather abundant among eelgrass and algæ at Noank and Woods-Holl, and probably other localities on the southern shore of New England. I formerly considered it as identical with *Tanais filum* Stimpson and supposed its range to extend as far as the Bay of Fundy. I now regard that as an error, as it is probable that *T. filum* is a true *Tanais* with simple uropoda, though I have as yet seen no specimens from the Bay of Fundy, nor any fully answering to Stimpson's description.

Paratanais limicola, sp. nov.

This species considerably resembles the preceding, but may be distinguished from it by the following characters: The eyes are small and rather inconspicuous, at least in the females, being only about half the vertical diameter of the antennulæ. The antennulæ have the second segment short, about half as long as the third. The dactylus of the second pair of legs, with its slender, acicular, terminal spine is longer than the propodus. The pleon is not dilated at the sides. The uropoda have the outer ramus two-jointed, slender, and surpassing the basal segment of the inner ramus which is five-jointed, with the basal segment long and imperfectly divided. Length 2·5mm.

This species was obtained on a soft muddy bottom in forty-eight fathoms, Massachusetts Bay, off Salem, in the summer of 1877, by the United States Fish Commission.

Paratanais cæca, sp. nov.

Body slender, elongated and loosely articulated. Head narrow in front, not broader than the bases of the antennulæ. Eyes wanting. Antennulæ four-jointed; first segment forming less than half its length; second segment longer than the third; last segment about as long as the second, slender, tapering and tipped with setæ. Antennæ attaining the tip of the third antennular segment. First pair of legs slender as compared with those of the preceding species, attaining the tip of the antennæ, basal segment subquadrate, hand or propodus less robust than the carpus; digital process of propodus serrated; dactylus short. Second (first free) thoracic segment two-thirds as long as the third, which is equal to the fourth and fifth; sixth and seventh progressively shorter. Second pair of legs scarcely more slender than the following pairs, basal segment not curving around the basal segments of the first pair. Pleon six-jointed; uropoda short, biramous, each ramus two-jointed, the outer more slender than the inner, half its length and bearing a long bristle at the tip. Length 2·5mm.

This species was taken along with *P. limicola* and unfortunately only a single specimen is as yet known.

Yale College, April, 1878.

NOTES ON NEW ENGLAND ISOPODA.

By OSCAR HARGER.

The marine Isopoda collected by the United States Commission of Fish and Fisheries having been placed in my hands by Professor Verrill, a report has been prepared including full descriptions, with figures of most of the species, except the *Bopyridæ*. Besides the collections of the Fish Commission, I have, through the kindness of Professor Verrill, had access to other extensive collections made principally by himself and Prof. S. I. Smith, at various points along the coast from Great Egg Harbor, New Jersey, to the Bay of Fundy, as is more fully detailed in the report now ready for publication. On account of unexpected delay in the publication of the report, it has been thought best to prepare the following brief summary of its contents, with especial reference to facts not hitherto published. Only such references are here given as are necessary to the understanding of the names adopted, and, in general, the distribution on the New England coast only is indicated.

The *Bopyridæ* have been identified by Professor S. I. Smith, who has also rendered other important assistance in the preparation of the report, of which the present paper may be regarded as an abstract.

The *Oniscidæ*, not being properly marine, are in general not included in the report; but three species, two of them as yet found only on the coast, are included as being commonly found by marine collectors. They are the first three of the following list, which embraces also all the marine Isopoda known to inhabit the waters of New England:

Philoscia vittata Say, Jour. Acad. Nat. Sci. Phil., vol. i, p. 429, 1818.

A southern species found as far north as Barnstable, Mass.

Scyphacella arenicola Smith, Rep. U. S. Fish Com., part i, p. 565 (271), 1874.

Sandy beaches, from Great Egg Harbor, New Jersey, to Nantucket, Mass. Not yet found north of Cape Cod.

Actoniscus ellipticus Harger, Am. Jour. Sci., III, vol. xv, p. 373, 1878.

Shores of Long Island Sound at Savin Rock, and Stony Creek, near New Haven. Collected by Professor Verrill.

Cepon distortus Leidy, Jour. Acad. Nat. Sci. Phil., II, vol. iii, p. 170, pl. xi, figs. 26-32, 1855.

"Branchial cavity of *Gelasimus pugilator*, Atlantic City, New Jersey."

Gyge Hippolytes Bate and Westwood, Brit. Sess. Crust., vol. ii, p. 230, 1868.—*Bopyrus Hippolytes* Kröyer, Grönlands Amfipoder, p. 306, pl. iv, fig. 22, "1838."

Parasitic on *Hippolyte*, etc., and found as far south as Massachusetts Bay.

Phryxus abdominalis Lilljeborg, Öfversigt af Kongl. Vetenskaps Akademiens Förh. Stockholm, 1852, p. 11.—*Bopyrus abdominalis* Kröyer, Naturhist. Tidssk., Bind iii. p. 102, 289, pl. 1, 2, (1840); Gaimard's Voyage en Scandinavie, etc., Atlas, pl. xxix, fig. 1 *a-u*, "1849."

Parasitic on *Pandalus*, *Hippolyte*, etc., and found as far south as Massachusetts Bay.

Dajus mysidis Kröyer, Gaimard's Voyage en Scandinavie, etc., Atlas, pl. xxviii, fig. 1, "1849."—*Bopyrus mysidum* Packard, Mem. Soc. Nat. Hist. Boston, vol. 1, p. 295, pl. viii, fig. 5, 1867.

Parasitic on *Mysis*, but not hitherto found south of Labrador.

Jæra albifrons Leach, Edinburgh Encyclopædia, vol. vii, p. 434, "1813-14"; Trans. Linn. Soc. London, vol. xi, p. 373, 1815.—*Jæra copiosa* Stimpson, Mar. Invert. Grand Manan. p. 40, pl. iii, fig. 29, 1853.

Common throughout the New England coast under sea-weed, in tide pools, etc. A comparison of specimens received from Oban, Scotland, through the kindness of Rev. A. M. Norman, indicates that our species must be regarded as identical with the well-known British species, and is therefore common to the two coasts.

Janira alta = *Asellodes alta* Stimpson, Mar. Invert. Grand Manan, p. 41, pl. iii, fig. 30, 1853.

A northern species not as yet found south of Massachusetts Bay, occasionally collected in tide-pools, but usually dredged, and extending to a depth of 190 fathoms.

This species is easily distinguished specifically from *J. maculosa* Leach, the type of the genus, but does not appear to differ by characters of generic importance, and I have therefore referred it to the older genus.

Janira spinosa, n. sp.

A second species of this genus was obtained in the summer of 1878, and on examination it appears to be as yet undescribed, although somewhat resembling *J. laciniata* G. O. Sars, but distinguished by the double instead of single row of spines along the dorsal region of the thorax.

The head is strongly rostrate, and has the antero-lateral angles acutely produced, but shorter than the median rostrum. The eyes are small and black, and placed a little behind the middle of the head, at about an equal distance from the median line and the lateral margin. The antennulæ are slender, and slightly surpass the first four segments of the antennæ. The antennæ are about as long as the head and thorax together, and the scale attached to the second peduncular segment is slender and pointed, surpassing the third segment. The flagellum forms about half the length of the antenna, and is slender, tapering, and multi-articulate.

The thoracic segments are all acutely produced at the sides into one or two salient angles, forming a row of acute serrations along the sides of the body. The first segment has a single angle produced somewhat forward around the sides of the head; the second, third, and fourth segments usually present two serrations, both the anterior and posterior angles being produced and acute, and the last three segments are produced and directed more and more backward. In the dorsal region, each segment bears a pair of sharp tubercles or spines. Anteriorly these spines are near the front margins of the segments and directed forward, but become posteriorly more erect and nearer the middle of the segment, and the last three pairs are directed backward, the last pair being near the hinder margin of the seventh segment. The legs are slightly spiny, the first pair but little thickened in the females. The pleon tapers at the sides, where it is minutely serrulate. Its posterior angles are salient and acute, like the anterior angles of the head. The uropods are of moderate length, about as long as the pleon, and composed of a cylindrical basal segment, bearing two rami, of which the inner is somewhat the larger, and nearly as long as the basal segment. Both, together with the basal segment, are sparingly bristly.

The color in alcohol is nearly white. Length 8mm.

Two specimens of this species were collected at Banquereau by Captain Collins, of the schooner Marion, August 25, 1878. They were found adhering to the cable of the schooner.

Munna Fabricii Kröyer, Naturhist. Tidssk., II. Bind ii, p. 280, 1-17; Gaimard's Voyage en Scandinavie, etc., Atlas, pl. 31, figs. 1 *a–q*, 1-29.

Casco Bay, near Portland, Me., Eastport and Western Bank, from low water to 150 fathoms.

Munnopsis typica M. Sars, Christiania Vidensk. Selsk., 1860, p. 84, 1861; Bidrag til Kundskab om Christiania Fjordens Fauna (Nyt Magazin), p. 70, pl. vi, vii, figs. 101-138, 1868.

This species has been taken in the Bay of Fundy in 60 fathoms; also, by Mr. J. F. Whiteaves, in the Gulf of Saint Lawrence.

Eurycope robusta Harger, Am. Jour. Sci., III, vol. xv, p. 375, 1878.

Not yet found south of the Gulf of Saint Lawrence, where it was taken by Mr. J. F. Whiteaves in 220 fathoms, muddy bottom.

Chiridotea cœca Harger, Am. Jour. Sci., III, vol. xv, p. 374, 1878.—*Idotea cæca* Say, Jour. Acad. Nat. Sci. Phil., vol. i, p. 424, 1818.

Common on the southern coast of New England, and taken as far north as Halifax in the summer of 1877.

Chiridotea Tuftsii Harger, Am. Jour. Sci., III, vol. xv, p. 374, 1878.—*Idotea Tuftsii* Stimpson, Mar. Invert. Grand Manan, p. 39, 1853.

This species has been taken at various points along the coast from Long Island Sound to Halifax, but was regarded as rare until the summer of 1878, when it was collected in abundance at Gloucester, Mass.

Idotea irrorata Edwards, Hist. nat. des Crust., tome iii, p. 132, 1840.—*Stenosoma irrorata* Say, Jour. Acad. Nat. Sci. Phil., vol. i, p. 423, 1818.—*Idotea tricuspidata* Desmarest, Dict. des Sci. nat., tome xxviii, p. 373, 1823; Consid. Crust., p. 289, 1825.

This species is common throughout the coast of New England, but is more abundant southward, being to a great extent replaced toward the north by the next species.

A comparison of English and European specimens with our own leaves no doubt of the identity of the species on the opposite coasts of the Atlantic. Being a common European species, it has been mentioned by many authors under a variety of names, which are more fully quoted and discussed in the report. Say's name appears to be the earliest that can be certainly connected with the species.

Idotea phosphorea Harger, Rep. U. S. Fish Com., part i, p. 569 (275), 1874.

Found throughout the coast, but more abundant northward.

Idotea robusta Kröyer, Naturhist. Tidssk., II, Bind ii, p. 108, 1846; Gaimard's Voyage en Scandinavie, etc., Atlas, pl. xxvi, fig. 3 a–r, 1849.

A pelagic species.

Synidotea nodulosa Harger, Am. Jour. Sci., III, vol. xv, p. 374, 1878.—*Idothea nodulosa* Kröyer, Naturhist. Tidssk., II, Bind ii, p. 109, 1846; Gaimard's Voyage en Scandinavie, etc., Atlas, pl. xxvi, fig. 2, 1849.

A northern species, found at Halifax, N. S., and 125 miles southward, in from 16 to 190 fathoms. Also from George's Bank.

Synidotea bicuspida = *Idotea bicuspida* Owen, Voyage of the Blossom, Crustacea, p. 92, pl. xxvii, fig. 6, 1839.—*Idotea marmorata* Packard, Mem. Soc. Nat. Hist. Boston, vol. i, p. 286, pl. viii, fig. 6, 1867.—*Idotea pulchra* Lockington, Proc. Cal. Acad. Sci., vol. vii, p. 45, 1877.

The determination of the synonymy of this species rests principally upon the work of Messrs. Streets and Kingsley in the Bulletin of the Essex Institute, vol. ix, p. 108, 1877. It has not yet been found south of the Grand Bank.

Erichsonia filiformis Harger, Rep. U. S. Fish Com., part i, p. 570 (276), pl. vi, fig. 26, 1874.—*Stenosoma filiformis* Say, Jour. Acad. Nat. Sci. Phil., vol. i, p. 424, 1818.

A southern species, not yet found north of Cape Cod.

Erichsonia attenuata Harger, Rep. U. S. Fish Com., part i, p. 570 (276), pl. vi, fig. 27, 1874.

Great Egg Harbor, New Jersey, and Noank, Conn. The species will probably be found at other localities, among eel-grass, on the southern shore of New England.

Epelys trilobus Smith, Rep. U. S. Fish Com., part i, p. 571 (277), pl. vi, fig. 28, 1874.—*Idotea triloba* Say, Jour. Acad. Nat. Sci. Phil., vol. i, p. 425, 1818.

A southern species, rare north of Cape Cod, but extending, with some other southern species, to Quahog Bay, on the coast of Maine.

Epelys montosus Harger, Rep. U. S. Fish Com., part i, p. 571 (277), 1874.—*Idotea montosa* Stimpson, Mar. Invert. Grand Manan, p. 40, 1853.

Replaces the preceding species for the most part at the north, but found also as far south as Long Island Sound. It has been obtained from a depth of 40 fathoms.

Astacilla granulata = *Leachia granulata* G. O. Sars, Arch. Math. og Naturvid. Christiania, B. ii, p. 351 (proper paging 254), 1877.—*Astacilla americana* Harger, Am. Jour. Sci., III, vol. xv, p. 374, 1878.

St. George's Banks, 1877, and Banquereau, 1878. I have seen no specimens of Sars's species for comparison, but his description appears to apply perfectly to the specimens described by myself before seeing his paper.

Sphæroma quadridentatum Say, Jour. Acad. Nat. Sci. Phil., vol. i, p. 400, 1818.

A southern species, scarcely passing north of Cape Cod, but occurring at Provincetown, Mass.

Limnoria lignorum White, Pop. Hist. Brit. Crust., p. 237, 1857.—"*Cymothoa lignorum* Rathke, Skrivt. af Naturh. Selsk. v. 101, t. 3, f. 14, 1799."—*Limnoria terebrans* Leach, Edinburgh Encyc., vol. vii, p. "433, 1813-14"; Trans. Linn. Soc. London, vol. xi, p. 371, 1815.

This genus was associated with the *Asellidæ* by Edwards without an examination of the specimens, and, so far as I know, he has been followed by recent authors. An examination of its structure appears to point unmistakably to affinity with the *Sphæromidæ*. I have not, however, thought best to include it in that family, but have placed it in a family by itself, the *Limnoriidæ*.

The species extends throughout the New England coast.

Cirolana concharum = *Conilera concharum* Harger, Rep. U. S. Fish Com., part i, p. 572 (278), 1874.—*Ega concharum* Stimpson, Mar. Invert. Grand Manan, p. 42, 1853.

Not found north of Cape Cod, but abundant at Vineyard Sound.

Cirolana polita = *Conilera polita* Harger, in Smith and Harger, Trans. Conn. Acad., vol. iii, p. 3, 1874.—*Ega polita* Stimpson, Mar. Invert. Grand Manan, p. 41, 1853.

St. George's Banks, Salem, and Eastport (Stimpson), rare.

Æga psora Kröyer, Grönlands Amfipoder, p. 318, "1838."—*Oniscus psora* Linné, Syst. Nat., ed. x, tom. i, p. 635, 1758.—*Ega emarginata* Leach, Trans. Linn. Soc. London, vol. xi, p. 370, 1815.

Parasitic on the Cod, Halibut, etc.; also dredged on St. George's Banks.

Nerocila munda Harger, Rep. U. S. Fish Com., part i, p. 571 (277), 1874.

On dorsal fin of *Ceratacanthus aurantiacus*, Vineyard Sound.

Ægathoa loliginea Harger, Am. Jour. Sci., III, vol. xv, p. 376, 1878.

Mouth of Squid, New Haven, Conn.

Livoneca ovalis White, List Crust. Brit. Mus., p. 109, 1847.—*Cymothoa ovalis* Say, Jour. Acad. Nat. Sci. Phil., vol. i, p. 394, 1818.

White and several other British carcinologists use the orthography *Lironeca;* but in the Dictionnaire des Sciences naturelles, tome xii, where the genus is established by Dr. Leach, the name occurs, in French and Latin, nine times on pages 352 and 353, spelled always with *v* as the third letter. I have, therefore, adhered to that orthography, although there is reason for supposing that Dr. Leach intended to use the form *Lironeca.*

Parasitic on Bluefish, etc.; not yet found north of Cape Cod.

Anthura polita Stimpson, Proc. Acad. Nat. Sci. Phil., vol. vii, p. 393, 1855.—*Anthura brunnea* Harger, Rep. U. S. Fish Com., part i, p. 572 (278), 1874.

A southern species, not found north of Cape Cod until the summer of 1878, when it was taken at Gloucester, Mass. Usually found among Eel-grass or mud in shallow water.

Paranthura brachiata = *Anthura brachiata* Stimpson, Mar. Invert. Grand Manan, p. 43, 1853.

A northern species, but found as far south as Vineyard Sound, from 27 to 115 fathoms.

Ptilanthura tenuis Harger, Am. Jour. Sci., III, vol. xv, p. 377, 1878.

Rare, but found throughout the New England coast. The remarkably elongate flagellum of the antennulæ belongs to the males only.

Gnathia cerina = *Praniza cerina* Stimpson, Mar. Invert. Grand Manan, p. 42, pl. iii, fig. 34, 1853; and, also, *Anceus Americanus* Stimpson, op. cit., p. 42, 1853; the former being the female form and the latter that of the adult male.

A northern species, not yet found south of Cape Cod, occurring in from 10 to 220 fathoms, and, in the young stages, parasitic on fish.

Tanais vittatus Lilljeborg, Bidrag til Känn. Crust. Tanoid., p. 29, 1865.—*Crossurus vittatus* Rathke, Fauna Norwegens (Nova Acta Acad., vol. xx,) p. 39, pl. i, figs. 1–7, 1843.

This species has been found at Noank Harbor, Conn., and will probably be found at other localities on our coast. I have had no European specimens for comparison, and, unfortunately, have not had access to some important European literature on the subject, but do not know of any character by which to distinguish it from Rathke's species, and have therefore regarded it as identical.

This genus is well separated from the next by the pleon, which bears only three pairs of pleopods and uniramous uropods, and by the remarkable incubatory sacs attached to the fifth thoracic segment of the females, and unlike anything else found among the *Isopoda.* They have been described by Rathke, Willemoes-Suhm, and others.

Leptochelia algicola = *Paratanais algicola* Harger, Am. Jour. Sci., III, vol. xv, p. 377, 1878.—*Leptochelia Edwardsii* Bate and Westwood, Brit. Sess. Crust., vol. ii, p. 134, 1868, (males).—*Tanais filum* Harger, Rep. U. S. Fish Com., part i, p. 573 (279), 1874, not of Stimpson.

A male specimen, received from Guernsey, through the kindness of

Rev. A. M. Norman, appears to agree perfectly with the males of this species, though not with Kröyer's description of *Tanais Edwardsii*. I have not therefore united my species with his, though I think it possible they may prove identical.

The species occurs in considerable abundance at Noank Harbor, Conn., among algæ, and also at Vineyard Sound, and will probably be found at other localities on the southern shore of New England. It has also been collected by Professor Verrill, during the present summer, at Provincetown, Mass., in company with *Limnoria* and *Chelura*, in old piles.

The genus *Leptochelia* has several years' priority over *Paratanais*, and, though founded on the male sex, ought, as I think, to be retained.

Leptochelia limicola = *Paratanais limicola* Harger, Am. Jour. Sci., III. vol. xv, p. 378, 1878.

Massachusetts Bay, off Salem, 48 fathoms, mud.

Leptochelia rapax, n. s.

Females of this species considerably resemble those of *L. limicola*, but may be distinguished by the following characters: The eyes are larger and more conspicuous; the last segment of the antennulæ is scarcely longer than the preceding, instead of nearly twice as long, as in *L. limicola*; the dactylus of the second pair of legs is somewhat shorter and the terminal spine less attenuated, and the external ramus of the uropods consists of a single very short and small segment, shorter than the basal segment of the inner ramus, which is not elongated. The inner ramus is five-jointed, instead of six-jointed, as in *L. algicola*.

The males are remarkable for the long and slender prehensile hand terminating the first pair of legs. The body of the males is short and robust, with the segments well marked by constrictions. The head, with the united first thoracic segment, is short and rounded, bulging strongly at the sides just behind the eyes, which are conspicuous, somewhat less in diameter than the bases of the antennulæ, distinctly articulated, and coarsely faceted. The antennulæ are elongated, especially in the basal segment, which is more than one-third as long as the body, slightly swollen on the inner side, near the base, then tapering to the tip; the second segment is cylindrical, less than half as long as and more slender than the first; the third is less than half the length of the second, and is followed by about eight short flagellar segments, the last one tipped with setæ. The antennæ, when extended, do not attain the end of the basal antennular segment; the first three segments are short, the fourth longest, being longer than the first three together, the fifth slender and tipped with setæ. The terminal setæ of both antennulæ and antennæ arise in part from minute or rudimentary terminal segments. The first pair of legs forms the most striking feature of this species. These legs, when extended, are in general longer than the body of the animal, though they vary considerably in size, being usually proportionally smaller in the smaller specimens. In these legs, the segments preceding the carpus are short and robust; but the carpus is about half as long as the body, and the propodus is even somewhat longer than the carpus, and usually strongly flexed beneath it. More than half the length of the propodus is made up of the slender digital process, which bears a low tooth on the inner side, near the base, and a stouter one near the slender incurved tip. The dactylus is slender, curved, and pointed, and armed with a few weak spinules along the inner margin. The forceps thus formed are capable of seizing and closing around the body of another individual.

The thoracic segments, except the first, are well separated; the second (first free) segment is shortest; the third, fourth, and fifth segments are of increasing length; the sixth is as long as the fifth; the seventh shorter. The first five segments of the pleon are of about equal length; the sixth shorter and obtusely pointed in the middle. The uropods consist on each side of a robust basal segment, bearing two rami, the outer short, and composed of a single segment, the inner five-jointed and tapering. Both rami are sparingly bristly. The males vary in length from 2.6^{mm} to 3.8^{mm}, and in breadth from 0.6^{mm} to 0.8^{mm}. The females are more slender. Color in alcohol nearly white or marked in the males by a brownish transverse band along the posterior margin of each segment.

This species was collected by Professor Hyatt and Messrs. Van Vleck and Gardner at Annisquam, Mass., in the summer of 1878.

Leptochelia filum = *Tanais filum* Stimpson, Mar. Invert. Grand Manan, p. 43, 1853.

"Bay of Fundy," Stimpson.

Leptochelia cœca = *Paratanais cœca* Harger, Am. Jour. Sci., III. vol. xv, p. 378, 1878.

Collected along with *L. limicola* in 48 fathoms, mud, Massachusetts Bay, off Salem, 1877.

Of the forty-three species enumerated in the preceding list, the following eighteen have as yet been found only north of Cape Cod:

Gyge Hippolytes Bate and Westwood.	*Synidotea bicuspida* Harger.
Phryxus abdominalis Lilljeborg.	*Astacilla granulata* Harger.
Dajus mysidis Kröyer.	*Cirolana polita* Harger.
Janira alta Harger.	*Ega psora* Kröyer.
Janira spinosa Harger.	*Gnathia cerina* Harger.
Munna Fabricii Kröyer.	*Leptochelia limicola* Harger.
Munnopsis typica M. Sars.	*Leptochelia rapax* Harger.
Eurycope robusta Harger.	*Leptochelia filum* Harger.
Synidotea nodulosa Harger.	*Leptochelia cœca* Harger.

The following ten have been found only south of Cape Cod:

Scyphacella arenicola Smith.	*Cirolana concharum* Harger.
Actoniscus ellipticus Harger.	*Nerocila munda* Harger.
Cepon distortus Leidy.	*Egathoa loliginea* Harger.
Erichsonia filiformis Harger.	*Livoneca ovalis* White.
Erichsonia attenuata Harger.	*Tanais vittatus* Lilljeborg.

The following fifteen have been found both north and south of Cape Cod:

Philoscia vittata Say.
Jæra albifrons Leach.
Chiridotea cæca Harger.
Chiridotea Tuftsii Harger.
Idotea irrorata Edwards.
Idotea phosphorea Harger.
Idotea robusta Kröyer.
Epelys trilobus Smith.

Epelys montosus Harger.
Sphæroma quadridentatum Say.
Limnoria lignorum White.
Anthura polita Stimpson.
Paranthura brachiata Harger.
Ptilanthura tenuis Harger.
Leptochelia algicola Harger.

The following eleven species occur also on the coast of Europe:

Gyge Hippolytes Bate and Westwood.
Phryxus abdominalis Lilljeborg.
Jæra albifrons Leach.
Munna Fabricii Kröyer.
Munnopsis typica M. Sars.
Idotea irrorata Edwards.

Astacilla granulata Harger.
Limnoria lignorum White.
Ega psora Kröyer.
Tanais vittatus Lilljeborg.
Leptochelia algicola Harger.

NOTICE OF RECENT ADDITIONS TO THE MARINE INVERTEBRATA, OF THE NORTHEASTERN COAST OF AMERICA, WITH DESCRIPTIONS OF NEW GENERA AND SPECIES AND CRITICAL REMARKS ON OTHERS.

PART I —ANNELIDA, GEPHYRÆA, NEMERTINA, NEMATODA, POLYZOA, TUNICATA, MOLLUSCA, ANTHOZOA, ECHINODERMATA, PORIFERA.

By A. E. VERRILL.

Among the very extensive collections made during the past eight years by the U. S. Commission of Fish and Fisheries, under the direction of Professor Baird, there are still many species not recorded as American in any of the reports hitherto published; most of these are well-known Arctic or Northern European species, but others are still undescribed. As the final reports on the different groups will require a long time for their completion, owing to the vast number of specimens to be examined from more than a thousand localities, it has been thought desirable to record some of the more important additions to the fauna, without further delay.* More detailed descriptions and numerous figures will be published in the final reports, together with the details of their geographical distribution. All the species included in the following list, unless otherwise stated, have been collected by the U. S. Fish Commission.

---

*Many species have also been recorded in various articles in the American Journal of Science and Arts, during several years past. See, also, an important paper on the Podophthalmous Crustacea, by Professor S. I. Smith, and one on the Pycnogonida, by E. B. Wilson, in the Trans. Conn. Academy, vol. v, 1879.

ANNELIDA.

Sthenelais gracilis, sp. nov.

A small, slender, delicate species. Scales white, smooth, outer edge with few (12-16) very small, unequal, tapering papillæ, which are not crowded, the longest about as long as the intervening spaces. Head short, broad, the posterior and lateral margins rounded, the front emarginate. Eyes black, conspicuous; the posterior pair on the dorsal surface in advance of the middle of the head; anterior pair nearer together, close to the anterior margin; median antenna long, stout at base, tapering to a slender tip; the palpi have about the same form and length as the median antenna. Dorsal setæ longer than the ventral, extremely slender, tapering gradually toward the very fine tips, and very minutely serrulate. Upper ventral setæ (2-4) simple, very slender, with the shaft smooth, the serrate portion broader, with rather long ascending spinules, the tips tapering to a long fine point; the median setæ, above the acicula, have longer, much stouter, smooth shafts, expanded distally, with the terminal portion long, curved, divided into eight to twelve imperfect joints, tapering to very slender capillary tips, which are mostly acute, sometimes faintly hooked. Below the acicula there are others, similar in structure, but with the shaft not so stout, and with the terminal piece shorter, with fewer joints; the lower portion of the fascicle consists of numerous, much more slender, capillary setæ, with smooth shafts and very long, slender, tapering, terminal pieces, composed of ten to twelve or more imperfect joints.

Harbor of Gloucester, Mass., 7 to 10 fathoms, sand, 1879 (U. S. Fish Commission). Described from alcoholic specimens.

Sthenelais Emertoni, sp. nov.

A small, slender species, with white, translucent scales, their outer edge with very small, nearly equal, slender papillæ, often slightly clavate at tip, and rather near together, their interspaces being mostly less than their length; surface partially covered with minute rounded verrucæ.

Dorsal setæ very slender, capillary, very minutely transversely serrulate. Few (about 4) upper ventral setæ, simple, long, slender, with the terminal portion sharply serrulate, the tips fine and sharp; next to these are some slender compound setæ, the terminal piece slender, straight, of moderate length, acute, with six to eight imperfect joints; the median setæ have much stouter, smooth shafts, expanded distally, and a nearly straight, short, rapidly tapering, sharply pointed, terminal piece, of four to six joints; below these are some with similar though smaller shafts, and a short, stout, terminal piece, hooked at the tip, and with a sharp ascending spine at about the distal third; others of the same size have the terminal piece very acute, with six to eight or more joints; the lowest are very slender, with a longer, very fine, tapering, terminal piece, imperfectly divided into about four to six joints, at each of which there is a projecting acute angle like a tooth; the last of

Bulletin of the Museum of Comparative Zoology,
AT HARVARD COLLEGE.
VOL. XI. No. 4.

REPORTS ON THE RESULTS OF DREDGING, UNDER THE
SUPERVISION OF ALEXANDER AGASSIZ, ON THE EAST
COAST OF THE UNITED STATES, DURING THE SUMMER
OF 1880, BY THE U. S. COAST SURVEY STEAMER "BLAKE,"
COMMANDER J. R. BARTLETT, U. S. N., COMMANDING.

(Published by permission of CARLILE P. PATTERSON and J. E.
HILGARD, Supts. U. S. Coast and Geodetic Survey.)

XXIII. — REPORT ON THE ISOPODA.

BY OSCAR HARGER.

CAMBRIDGE:
PRINTED FOR THE MUSEUM.
SEPTEMBER, 1883.

No. 4. — *Reports on the Results of Dredging, under the Supervision of* ALEXANDER AGASSIZ, *on the East Coast of the United States, during the Summer of* 1880, *by the U. S. Coast Survey Steamer* "*Blake*," COMMANDER J. R. BARTLETT, U. S. N., Commanding.

(Published by permission of J. E. HILGARD, Superintendent of the U. S. Coast and Geodetic Survey.)

XXIII.

Report on the Isopoda. By OSCAR HARGER.

THE collection of Isopoda from the Blake Expedition, although small in number, is remarkable for the large proportion of interesting forms secured, since nearly all the specimens prove to belong to species that are either new, or not hitherto known from our coast, or to species known only from single specimens and hence only imperfectly described.

CIROLANIDÆ.

Cirolana spinipes BATE & WESTWOOD.

Plate I. Figs. 2-2 d. Plate II. Figs. 1-1 c.

Cirolana spinipes BATE & WESTWOOD, Brit. Sess. Crust., II., p. 299. 1868.

Specimens of this species, not hitherto recorded from our coast, were obtained from two localities; viz. Station 316, Lat. 32° 7′ N., Long. 78° 37′ 30″ W., 229 fathoms, one female; and Station 321, Lat. 32° 43′ 25″ N., Long. 77° 20′ 30″ W., 233 fathoms, three females and one male.

These specimens appear to agree perfectly in all specific characters with others in the collection of the Yale College Museum identified and sent to the Museum by the Rev. A. M. Norman, from the Shetland Islands. They do, however, differ in some respects from the description of that species in Bate and Westwood's work, and to facilitate comparison with that species and with others on our coast a full description is appended, with figures.

The body is a little more than three times as long as broad, with the dorsal surface strongly rounded, polished and smooth except for minute punctations, mostly near the posterior margin of each segment, and a median dorsal row of shallow oval depressions, most distinct on the third, fourth, and fifth thoracic segments.

The head is quadrate, widest across the posterior part of the eyes, which are oval, and more distinct than in the other species on our coast. A horizontal impressed line passes along the side of the head above and in front of the eye, and another just above the anterior margin over the bases of the antennæ. The antennulæ (Pl. I. Fig. 2 a) are short, not equalling the peduncle of the antennæ. Their basal segments are in contact above and in front; the second segment is short, the third as long as the first two, and is followed by a flagellum not as long as the peduncle and composed of about fourteen short and closely united segments. The antennæ (Pl. I. Fig. 2 b), when reflexed, reach the posterior margin of the third thoracic segment; the first two peduncular segments are short; the third and fourth each twice as long as the second, and of somewhat greater diameter; the fifth is the longest peduncular segment, and, at base, only about half the diameter of the fourth. The slender, tapering flagellum is about twice as long as the peduncle, and composed of twenty-five or more segments. The fourth and fifth peduncular segments bear, near their distal ends, a few slender and rather short bristles, much less conspicuous than in *C. concharum* or *C. polita* (Pl. I. Fig. 1 b), but longer than in the next species, *C. impressa* (Pl. I. Fig. 3 b).

The first thoracic segment is slightly longer than the second; posteriorly the segments increase slightly in length to the fifth or sixth, but the seventh is the shortest. The first segment is marked by an impressed curved line just above the lateral margin. The epimera of the second and third segments are small, subquadrate, rounded behind. The fourth epimeron is larger than the preceding ones, with the lower posterior angle rounded. The fifth and sixth epimera are of about equal size and larger than the others, while the seventh is the smallest of all. In the last three the posterior margin is oblique, and the lower angle is pointed. All the epimera are quadrate in general outline, and from near the middle of the line of union with the segment a sharp depressed line extends upward upon each of the last four segments.

In the first pair of legs (Pl. II. Fig. 1 a) the basis is flattened on the upper or inner side, and slightly curved in adaptation to the convex under surface of the head. The anterior margin of this segment is also fringed with bristly hairs. The succeeding segments are well armed with bristles, and the merus, carpus, and dactylus are armed along their palmar margins also with acute spines; the carpus in this leg is triangular and articulated with little motion to the propodus. The legs of the second and third pairs resemble the first, but have a free articulation between the propodus and carpus, which is oval and armed with several acute spines. These three pairs of legs are directed forward. The fourth and subsequent pairs are directed backward. The legs of the fourth pair (Pl. II. Fig. 1 b) are of moderate length and well armed with bristles or spines throughout, especially on the merus and carpus, where the spines form a striking feature. The palmar margin of both these segments is armed with a row of slender elongated spines and bristles, with many shorter spines also along the margin, while upon the outer or exposed surface of both segments is a pretty regular longitudinal row of short spines, nearly along the middle of the segment, and others in less regular order between this row of spines and the palmar border. In the English specimens these spines are even somewhat more numerous than in ours. The fifth leg is similar to the fourth, but somewhat longer and more slender, and the spines on the merus and carpus are nearly as pronounced and definitely arranged as in the fourth, while a similar arrangement is found in a less degree upon the remaining two pairs of legs. In the last two pairs of legs the bases are flattened, expanded, and well ciliated, forming strong swimming organs. One of the last pair is figured on Plate II. Fig. 1 c.

All of the pleonal segments are plainly evident above, the first not being at all concealed by the last thoracic segment, as in the other species on our coast. The first four segments are subequal in length on the median dorsal line; laterally they are carinated, the carina ending behind in an angulation (see Pl. I. Fig. 2 c) which is most pronounced on the third segment and is rounded off on the fourth. The thickened, chitinous walls of these segments are more or less continued below the lateral keel upon the inferior surface of the pleon, and in the first two segments the inner and posterior angles of this portion are acutely produced, in the second segment, into short, divergent spiniform processes. In the third, the under part of the segment runs out to its lateral angle, and in the fourth segment this portion is small and not angulated. All these segments are smooth and not ciliated laterally. The fifth segment is small, and does not reach the lateral margin of the pleon. The last segment (Pl. II. Fig. 1) is semioval, acutish at the tip, near which it is ciliated and bears a few short spines. The basal segment of the uropod is produced at the inner angle to about half the length of the outer ramus. This ramus is lanceolate in outline, shorter than the inner, and of only about half its width; both are ciliated and armed with short spinules. The inner is destitute of the emargination seen on the outer border near the tip in the other species. The second pair of pleopods in the male (Pl. I. Fig. 2 d) is armed, on its inner ramus, with a stylet of peculiar form. The stylet is slightly longer than the ramus and very acute at the tip, just below which it is suddenly much expanded and sends off a prong on the outer side, toward the lamella, as shown in the figure. A similar structure is seen in the male from the Shetland Islands, but I have seen nothing like it in the other American species.

Length of female, 23 mm.; breadth, 7.5 mm. The single male specimen obtained is smaller; length, 16 mm.; breadth, 5.5 mm.

Cirolana impressa sp. nov.

Plate I. Figs. 3 - 3 d. Plate II. Figs. 3 - 3 c.

This species closely resembles *C. polita* (Stimp.), as may be seen from the figures of the two species (Pl. I. Fig. 1, *C. polita*, Fig. 3, *C. impressa*). They are most readily distinguished by the impressed lines on the surface of the epimera in the present species, but a closer inspection brings to light other characters, as will appear in the following description.

The body is more than three times as long as broad, with the sides nearly straight and parallel, smooth and polished, with fewer punctations than in *C. polita*, but with the usual median dorsal row.

Head rounded hexagonal, broadest across the eyes, with an impressed line just above them extending around the front of the head. Eyes small, subtriangular, notched on their front outline by a thickened marginal ridge, which dies out in the ocular region. Antennulæ (Pl. I. Fig. 3 a) about as long as the peduncle of the antennæ ; two basal segments swollen and together longer than the third; flagellum as long as the peduncle, composed of about a dozen segments, shorter and more closely articulated than in *C. polita*. (Pl. I. Fig. 3 b). Antennæ surpassing the margin of the first segment, shorter than in the preceding species ; flagellum one half longer than the peduncle and composed of about twenty-two segments.

First thoracic segment closely adapted to the hinder margin of the head, about twice as long on the median line as the second. Behind the second, the segments gradually increase in length to the seventh, while in *C. polita* the fifth is the longest segment and the seventh is shorter than the sixth. The first segment is marked in the epimeral region by a nearly marginal impressed line. In the following segments the epimera are distinct and increase in size to the last. The second and third epimera are subquadrate, with rounded posterior angles, much as in *C. polita*, but each is marked by a curved impressed line below and somewhat behind the middle. The third and fourth epimera are also quadrate in outline, the posterior margins becoming oblique and meeting the inferior margin in each at an angle, while in *C. polita* both these epimera are rounded behind. In the present species, moreover, both these epimera are marked with an oblique impressed line running from near the middle of the upper margin toward the lower posterior angle. The last two epimera are subtriangular in outline, as in *C. polita*, and the sixth is marked with an impressed line, much as in the fourth and fifth. A similar line is faint, or represented by a row of punctations, on the last epimeron. The impressed lines on the epimera of this species serve also to distinguish it from *C. concharum* (Stimp.), to which it has considerable resemblance.

In the first pair of legs (Pl. II. Fig. 3 a) the merus is large and produced at its outer angle beyond the middle of the propodus, its palmar margin is armed with acute spinules much as in *C. polita*, but not quite as strong as in that species (Pl. II. Fig. 2 a), while it differs from *C. concharum* (Pl. II. Fig. 4 a) in lacking the row of blunt spinules near the palmar margin of this segment. The legs of the fourth pair (Pl. II. Fig. 3 b) are armed with spines, with comparatively few bristles among them, and the spines upon the surface of the merus and carpus are arranged transversely, instead of as in the last species. In the seventh pair of legs (Pl. II. Fig. 3 c) the basis is slender and nearly naked, as in *C. concharum* (Pl. II. Fig. 4 c), and the three following segments are flattened and furnished with close-set bristles distally.

The pleon (Pl. I. Fig. 3 c) is more overlapped and concealed by the last thoracic segment than in either *C. concharum* (Pl. I. Fig. 4) or *C. polita* (Pl. I.

Fig. 1 c). The first segment is quite concealed above, and the second more or less concealed also in the ordinary position of the segments. In the ventral portions of the first three pleonal segments the posterior angles are rounded instead of being acute, as in both the *C. polita* and *C. concharum* ; laterally the second, third, and fourth segments are ciliated, as in both those species. The telson (Pl. II. Fig. 3) is much like that of *C. polita* (Pl. II. Fig. 2). The uropods have the basal segment produced internally; the outer ramus is about half as wide as the inner, which has a distinct notch near the distal end of the outer border and is obliquely truncate, or, in the larger specimens, emarginate at the end; both rami, like the end of the telson, are strongly ciliated, but sparingly spinulose. The telson is distinguished from that of *C. concharum* (Pl. II. Fig. 4) by the emargination at the tip in that species. The stylet on the second pair of pleopods in the male (Pl. I. Fig. 3) is simple, ensiform, and tapers to a blunt point ; it surpasses the lamella to which it is attached.

The four large females of this species obtained by the Blake Expedition measure in length 21–23 mm, and in breadth 6–6.5 mm. Specimens obtained by the U. S. Fish Commission are many of them smaller, but vary from 15 to 27 mm. in length.

The specimens were obtained at Station 336, Lat. 38° 21' 50" N., Long. 73° 32' W., from a depth of 197 fathoms. Others have also been obtained by the U. S. Fish Commission at the following stations: —

Station.	Fathoms.	N. Lat.	W. Long.	Specimens.
871	115	40° 2' 24"	70° 23' 40"	4
949	100	40° 3'	70° 31'	11
1094	301	39° 57'	69° 47'	1
1095	321	39° 55' 28"	69° 47'	2

ÆGIDÆ.

Æga psora (LINNÉ) KRÖYER.

One specimen from 306 fathoms at Station 303 in Lat. 41° 34' 30" N., Long. 65° 54' 30" W.

? Æga Webbii (GUÉRIN) SCHIÖDTE & MEINERT.

Pterclas Webbii GUÉRIN, Mag. Zoöl., Classe VII., Pl. XX. 1836.
Æga Webbii SCHIÖDTE & MEINERT, Naturhist. Tidssk., R. III., B. XII., p. 347. Pl. X. (Cym. IV.) Figs. 1–4. 1879.

A single immature specimen of this, or a closely allied species, was taken at Station 307, Lat. 31° 57' N., Long. 78° 18' 35" W., from a depth of 333 fathoms. It measures 10.5 mm. in length, 5.5 mm. in breadth, and has not yet developed the seventh pair of legs, but the propodi of the second and third pair of legs are armed with the characteristic cultriform spine, and I have referred it to this species, though not with certainty.

Æga incisa Schiödte & Meinert.

Plate III. Fig. 1.

Æga incisa Schiödte & Meinert, Naturhist. Tidssk., R. III., B. XII., p. 373, Pl. X. (Cym. IV.) Figs. 13–15. 1879.

A single specimen, apparently of this species, was taken at Station 307, from a depth of 333 fathoms, in Lat. 31° 57′ N., Long. 78° 18′ 35″ W.

It agrees so closely with Schiödte and Meinert's description that I have little doubt of its identity with that Mediterranean species, although the body is proportionally narrower and the segments of the pleon more regularly curved above than represented in the figure of *Æ. incisa* given by those authors.

In our specimen the body is nearly three times as long as broad, sparingly punctate, well rounded above.

The head is rounded behind, presenting no ocular lobes; in front it is produced into a distinct, pointed process projecting downward between the bases of the antennulæ, separating them and nearly touching the frontal lamina, which is small and rhomboidal. The first two segments of the antennulæ are short and small, and rounded in front, not enlarged as in *Æ. psora* Kröyer; the third segment is longer than the first two, and is followed by a slender flagellum, longer than the peduncle, composed of about fifteen segments, of which the first is the longest, being as long as the next two, instead of "quam secundo paulo longiore," as in the typical specimen of the species. The antennæ, when reflexed, surpass the second thoracic segment; the flagellum is longer than the peduncle, and composed of less than twenty segments.

The eyes are large, and meet broadly on the median line; ocelli in about ten horizontal rows, half of which meet on the median line in front.

The first thoracic segment is longer than the second, thence they increase slowly in length to the fifth or sixth, and the seventh is short. The epimeral region of the first segment is marked by an obliquely descending depressed line. The epimera are all angulated behind, though only the last two are sensibly produced, and all are marked by one or two oblique curved lines, running downward and backward, the posterior one ending in the lower angle. The last epimeron does not attain the lateral angle of the first segment of the pleon. The legs are weak, and armed with but few small and short spinules throughout.

All the segments of the pleon are evident, but the first is very short above; the first four are distinctly angulated laterally. The telson is subtriangular, distinctly notched behind, as well as minutely crenulated and spinulose. The basal segment of the uropods is produced internally about half the length of the inner ramus, which is obliquely elongate triangular, larger and broader than the narrowly ovate outer one; both are ciliated and minutely denticulate.

Length, 13.5 mm.; breadth, 5 mm.

I have seen no other specimens.

Rocinela oculata sp. nov.

Plate III. Figs. 2 - 2 a. Plate IV. Fig. 1.

Body oval, length a little more than twice the breadth, surface sparsely punctate.

Head subreniform, produced in front into a truncated process over the bases of the antennulæ, yoke-shaped behind, the ocular lobes projecting, upper surface nearly covered with the large eyes in which the ocelli are large and quincuncially arranged in ten rows along the long axis of each eye. Five of these rows meet along the median line.

The antennulæ are slender and scarcely attain the tip of the antennal peduncle; the basal segment is short and concealed from above; the second is longer than the first; the third is slender, but not as long as the first two together; flagellum about as long as the peduncle, slender and composed of five segments, of which the first is much the longest and the last is the shortest, and does not quite attain the posterior border of the eye when the antennula is reflexed. The antennæ surpass the first thoracic segment; the first two segments are very short; the flagellum is about twelve-jointed.

First thoracic segment closely adapted to the head in front; fourth segment longest on the median line above; sixth short; seventh nearly concealed and quite small, although bearing a well-developed pair of legs below.

The epimera of the second and third segments are oblique, but not acute nor produced backward in a lateral view; in the four following segments they are produced and very acute; the seventh epimeron is much smaller than the sixth, and, owing to the shortness of the seventh segment, ends behind about on a line with it, both epimera surpassing the first segment of the pleon.

Legs of the first pair (Pl. IV. Fig. 1) slender, armed with a long slender dactylus, much curved near its base; propodus expanded with a large palmar lobe armed with a marginal row of eight curved spines; carpus short, with a single curved palmar spine. Legs of the second and third pair much like the first, but with only six spines on the propodus. Legs of the fourth and posterior pairs slender, armed with spines principally at the distal ends of the ischium, merus, and carpus.

First segment of pleon very short and nearly concealed by the thoracic segments, narrower than the next three segments, which are about equal, acutely produced at the sides so as to resemble in shape the seventh epimeron; fifth segment narrower than fourth, but somewhat longer on the median line; telson semi-oval, regularly rounded behind and ciliated. Uropods equalling the telson; inner angle of basal segment produced, about one third the length of the inner ramus, which is ligulate, rounded behind, slightly shorter than the outer, and less than half as broad; outer ramus obovate, spinulose along the outer border; both rami ciliated except near the base.

Length, 13.5 mm.; breadth, 6 mm.

A single specimen of this species, the only one as yet known, was taken at Station 305, Lat. 32° 18′ 20″ N., Long. 78° 43′ W., from a depth of 252 fathoms.

Rocinela Americana Schiödte & Meinert.

Plate III. Figs. 3, 3 a, 4. Plate IV. Figs. 2, 2 a.

Rocinela Americana Schiödte & Meinert, Naturhist. Tidssk., R. III., B. XII., p. 394, Pl. X. (Cym. IV.) Figs. 16–18. 1879.

Two specimens of this species were obtained at Station 320, Lat. 32° 33′ 15″ N., Long. 77° 30′ 10″ W., from a depth of 257 fathoms, and a considerable number of other specimens obtained at various localities by the U. S. Fish Commission enable me to add somewhat to Schiödte and Meinert's description of the species, which was drawn from a single female specimen. A comparison of their type, from Trenton,* Maine, now preserved in the Museum of Comparative Zoology at Cambridge, and kindly loaned for the purpose by Professor Agassiz, shows no differences that can be regarded as specific.

The body is oval, with the length more than twice the breadth, and nearly all of our specimens are proportionally broader than the type, although none of them are quite as large.

Head subtriangular, rounded behind, acutish or slightly produced in front, more distinctly produced and somewhat angulated in front in the males (Pl. III. Fig. 4). Eyes rather large, separated by about one quarter the diameter of the head, rounded behind, more or less angulated at the point of nearest approach, where, in the males, a distinct angle of a hexagon is seen at the meeting of two rows of nine and six ocelli along the inner margin of the eye, one ocellus at the angle being common to both rows.

The antennulæ, when reflexed, only slightly surpass the head, and the flagellum is composed of five or six segments, of which the first is not much elongated and the last nearly attains the end of the antennal peduncle. The antennæ nearly attain the hinder margin of the second thoracic segment; the first and second segments are very short and concealed by the projecting front; the flagellum is as long as the peduncle, and composed of about fourteen segments.

The first thoracic segment is slightly excavated for the ocular lobes of the head; epimera of second and third segments subquadrate, oblique but not acute behind, marked with an impressed line near the lower margin; remaining four epimera acute and moderately produced; last epimeron usually surpassing the first segment of the pleon, although in some of the larger females, as in the type specimen, it fails to do so.

Prehensile legs (Pl. IV. Fig. 2) armed with three acute spines on the palmar margin of the propodus, and three obtuse spines on the same margin of the

* Trenton is incorrectly printed "Ireston" in Schiödte and Meinert's paper.

merus; carpus short. Ambulatory legs (Pl. IV. Fig. 2 a), well armed with spines.

First segment of pleon small, nearly concealed by the last thoracic segment, and usually surpassed by the last pair of epimera, narrower than the three following segments, which are slightly broader than the last thoracic segment without the epimera. Last segment broader than long, rounded and ciliated behind, faintly furrowed on the median line posteriorly. Uropods about equal to the telson; basal segment more or less produced at the internal angle, outer ramus shorter than the inner, both rounded behind and ciliated, denticulated externally, with short spinules in the notches between the teeth.

The female specimens vary in length from 14 mm. to 25 mm. and in breadth from 6 mm. to 10 mm., being mostly slightly broader in proportion than the type specimen, which is 26.5 mm. long, 10 mm. broad. The large male in the Blake Collection is 28 mm. long, 12 mm. broad; the small female, 17.5 mm. by 7 mm. A male collected by the U. S. Fish Commission at Station 871 is 22 mm. long, 9.5 mm. broad.

The typical specimen of this species is destitute of color markings, which may however have faded out from exposure to the light. Nearly all the other specimens are rather distinctly marked, chiefly along the sides of the body, with dark brown, arranged as follows. The lateral margins of the first thoracic segment, and the epimera sometimes of the third, and usually of the fourth, fifth, and sixth segments, but not of the seventh, are dark or nearly black, and the color extends distinctly to the adjacent regions of the fourth segment, and may extend across the back along the hinder margin of this segment; the next two segments may be similarly, but less strongly marked. On the pleon the color appears as a curved or crescentic band, along the lateral margins of the second, third, and fourth segments, and across the back part of the fifth and fore part of the sixth segments. On the sixth segment the color when present is divided by the median line into two more or less distinct spots, or maculæ. The posterior part of the telson is lighter-colored than the body.

This species has also been obtained by the U. S. Fish Commission at the following stations: —

Station.	Fathoms	N. Lat.	W. Long.	Specimens.
871	115	40° 2' 54"	70° 23' 40"	5
874	85	40° 0'	70° 57'	Cast skin.
875	126	39° 57'	70° 57' 30"	1
897	157	37° 25'	74° 18'	2
1108	101	40° 2'	70° 37' 30"	1
Oct. 4, 1882	Trawl-line			1

Rocinela sp.

A single specimen, probably of an undescribed species of this genus, was obtained at Station 344, Lat. 40° 1' N., Long. 70° 58' W., from 129 fathoms.

This specimen, although 27 mm. in length, is not yet adult, as shown by the rudimentary condition of the seventh pair of legs, and differs from the preceding especially in having the eyes more finely granulated. The material is too incomplete to attempt a full description.

Syscenus infelix Harger.

Plate III. Figs. 5, 5 a.   Plate IV. Figs. 3 - 3 h.

Syscenus infelix Harger, Rep. U. S. Fish Com., Pt. IV. for 1878, p. 387. 1880.

Three specimens of this species were obtained at two localities; viz. a single female at Station 303, Lat. 41° 34′ 30″ N., Long. 65° 54′ 30″ W., from 306 fathoms, and two males at Station 309, Lat. 40° 11′ 40″ N., Long. 68° 22′ W., from 304 fathoms. Besides these specimens a considerable number have also been obtained by the U. S. Fish Commission, from various localities along the coast as far south as Delaware Bay, and from a depth as great as 372 fathoms, so that the species, originally described from a single specimen, has now become comparatively common in the collection, and I am enabled to make some corrections in the description already given, as well as to add further details and present figures of the species.

Many of the specimens since obtained are larger than the type, and such examples often have the body quite distinctly corrugated and rather coarsely pitted, especially upon the head and the anterior part of the thorax or pereion. In some of the larger males the ocular regions on each side of the head are swollen and distinctly pitted and corrugated. On the lateral margin of the head is a notch, into which may be received a short process on the anterior angle of the first segment, thus producing a very firm articulation when the head is drawn closely against the first segment. The flagellum of the antennula is usually composed of seven segments instead of six, but the number may be different on opposite sides of the same specimen. A bottom view of the head, enlarged eight diameters, is given on Plate IV. Fig. 3, showing the antennary organs, the right antenna being removed to show the antennula of that side.

The maxillipeds (Pl. IV. Fig. 3 c) are robust, thickened along the inner or median side where they meet; the first segment of the palpus is large, nearly square, and armed at its inner distal angle with a single hook; its distal margin is shorter than the proximal, and is angulated at the articulation with the second short transverse segment. This segment is armed distally with three hooks, of which the anterior appears to be articulated and should perhaps be regarded as a third segment of the palpus. The outer or second maxillæ are thin, delicate, and obscurely lobed at the tip, where they are armed with a single small hook. The inner or first maxillæ (Pl. IV. Figs. 3 b, 3 b′) are armed with spines, of which the inner are shorter and straight, the outer are larger and curved or hooked at the tip. The mandibles (Pl. IV. Fig. 3 a) are robust at base, but slender and acute at the tip.

In the prehensile, or first three pairs of legs, the merus, carpus, and propodus are each armed with a short, curved, blunt spine on the palmar margin, as shown in the figure of a leg of the first pair on Plate IV. Fig. 3 d. The remaining four pairs of legs, not all natatory, are well fitted for prehension by their slender curved claws, and differ considerably in their proportions in specimens of different sizes, as shown by the accompanying table of measurements. All the legs are strongly flexed at the articulation of the basis with the ischium. In the sixth and seventh pairs, the ischium, merus, carpus, and propodus are elongated and in the small specimens slender, so that, with the addition of the dactylus, the last five segments of the leg of the sixth pair may attain to five sixths or even seven eighths the length of the body. The bases do not participate in this elongation and are therefore omitted in the measurements, since to include them would only diminish the contrast between the large and small specimens, shown especially in the last six columns of the table. In large specimens, like the one figured, the sixth and seventh pairs of legs are much more robust than in smaller ones.

The pleopods (Pl. IV. Fig. 3 g) are not naked, as originally described, but all the anterior ones, as usual in the *Ægidæ*, are distinctly ciliated. The cilia are however short and not very evident, and were overlooked in the single specimen described. In the small specimens they are proportionally longer than in larger ones. The second pair of pleopods in the male (Pl. IV. Fig. 3 g) bears a slender stylet tapering to the tip, and about as long as the ramus to which it is attached. In the small specimen, whose measurements are given in the last column of the table, the stylet is blunt, and considerably shorter than the ramus. The uropods (Pl. IV. Fig. 3 h) are robust; the basal segment is oblique, but not much produced internally; the rami are well ciliated.

Professor Verrill states that in life this species is bright colored, varying from bright orange to salmon-colored above and light yellow underneath. This color soon fades in alcohol.

Considerable variations in size, and corresponding variations in the proportions, especially of the sixth and seventh pairs of legs, are shown in the following table of measurements, in which the first three columns contain measurements of the Blake Expedition specimens, the next four columns contain measurements of specimens obtained at a single locality (Station 945) off Martha's Vineyard, by the U. S. Fish Commission in the summer of 1881, while in the last column are measurements of a smaller specimen obtained by the Fish Commission at another locality (Station 1028) in the same region. The measurements in the fourth column are from the specimen figured on Plate III. Figs. 5 and 5 a; those of the next five columns are from specimens gradually decreasing in size to the last. The length of the ambulatory legs, especially those of the sixth and seventh pairs, is seen to increase proportionally as the length of the body diminishes, except in the case of the seventh pair of legs of the last specimen. This is doubtless to be explained as a mark of immaturity in addition to the one already noted in the second pair of pleopods. The measurements are in millimeters, and the proportion of each to the length of the body is indicated by the accompanying decimal.

MEASUREMENTS.*

Syscenus infelix II.	B. ♀ 343	B. ♂ 509	B. ♂ 309	F. C. ♂ 945	F. C. ♂ 945	F. C. ♂ 945	F. C. ♀ 945	F. C. ♂ 945	F. C. ♂ 1028
Length of body	1.00 24.5	1.00 31.0	1.00 30.0	1.00 44.0	1.00 32.0	1.00 27.0	1.00 25.0	1.00 18.0	1.00 15.0
Transverse diameter of head	.16 3.8	.16 5.0	.15 4.5	.14 6.0	.16 5.0	.15 4.0	.16 4.0	.18 3.2	.19 2.8
" " 1st segment	.33 8.0	.29 9.0	.31 9.2	.24 12.5	.28 9.0	.30 8.0	.30 8.0	.33 6.0	.33 5.0
" " 3d segment	.36 8.8	.35 11.0	.35 10.5	.34 15.0	.33 10.5	.33 9.0	.36 9.0	.41 7.4	.35 5.2
" " 7th segment	.27 6.5	.26 8.0	.27 8.0	.26 11.5	.25 8.0	.27 6.2	.26 6.0	.25 4.5	.27 4.0
" " pleon at base	.20 5.0	.18 5.5	.18 5.5	.19 8.0	.17 5.5	.17 4.5	.20 5.0	.20 3.6	.21 3.0
" " last segment of pleon	.23 5.6	.23 7.2	.26 7.8	.25 11.2	.24 7.5	.21 5.6	.21 5.8	.22 4.0	.21 3.2
Longitudinal diameter of last segment of pleon	.23 5.6	.26 8.0	.29 8.6	.27 12.0	.26 8.2	.21 5.6	.26 5.8	.25 4.5	.27 4.0
Length, beyond basis, of leg of 3d pair	.18 4.5	.16 5.0	.17 5.0	.14 6.0	.18 5.2	.15 4.0	.17 4.2	.17 3.0	.20 3.0
" " " " 4th pair	.30 7.0	.25 8.0	.27 8.0	.25 11.0	.25 8.0	.26 7.0	.28 7.0	.28 5.0	.33 5.0
" " " " 5th pair	.37 9.0	.35 11.0	.33 10.0	.30 13.0	.31 10.0	.33 9.0	.34 8.5	.42 7.5	.40 6.0
" " " " 6th pair	.66 16.0	.58 18.0	.53 16.0	.45 20.0	.50 16.0	.62 17.0	.68 17.0	.82 15.0	.87 13.0
" " " " 7th pair	.53 13.0	.50 15.5	.47 14.0	.41 18.0	.42 13.5	.54 14.6	.60 15.0	.67 12.0	.61 9.2

* In the table of measurements B. is used to denote the Blake Expedition, F. C. the U. S. Fish Commission, and the accompanying numbers refer to the stations at which the specimens were obtained. The measurements are in millimeters, and over each is placed in small figures the corresponding decimal part of the length of the body.

NEW HAVEN, September 6, 1883.

EXPLANATION OF THE PLATES.

PLATE I.

Fig. 1. *Cirolana polita* Harger ex Stimpson. Lateral view of female, enlarged three diameters.
" 1 a. Antennula of another specimen, enlarged twelve diameters.
" 1 b. Antenna of same, enlarged twelve diameters.
" 1 c. Lateral view of pleon of *C. polita* as in fig. 4, enlarged five diameters.
" 2. *Cirolana spinipes* Bate & Westwood. Lateral view of female, enlarged three diameters.
" 2 a. Antennula of another specimen, enlarged ten diameters.
" 2 b. Antenna of same, enlarged ten diameters.
" 2 c. Pleon of *C. spinipes* as in fig. 4, enlarged five diameters.
" 2 d. Pleopod of the second pair of *C. spinipes*, male, enlarged eight diameters.
" 3. *Cirolana impressa* Harger. Lateral view of female, enlarged three diameters.
" 3 a. Antennula of another specimen, enlarged twelve diameters.
" 3 b. Antenna of same specimen, enlarged twelve diameters.
" 3 c. Pleon of *C. impressa* as in fig. 4, enlarged five diameters.
" 3 d. Pleopod of the second pair of *C. impressa*, male, enlarged eight diameters.
" 4. Pleon of *Cirolana concharum* Harger ex Stimpson, showing the first five segments in a lateral view, with dotted outline of last thoracic segment and its epimeron, enlarged five diameters.

PLATE II.

Fig. 1. *Cirolana spinipes* Bate & Westwood. Last segment of pleon with uropods, enlarged six diameters.
" 1 a. Leg of the first pair, enlarged eight diameters.
" 1 b. Leg of the fourth pair, enlarged six diameters.
" 1 c. Leg of the seventh pair, enlarged six diameters.
" 2. *Cirolana polita* Harger ex Stimpson. Last segment of pleon with uropods, enlarged six diameters.
" 2 a. Leg of the first pair, enlarged eight diameters.
" 2 b. Leg of the fourth pair, enlarged eight diameters.
" 3. *Cirolana impressa* Harger. Last segment of pleon with uropods, enlarged six diameters.

Fig. 3 a. Leg of the first pair, enlarged eight diameters.
" 3 b. Leg of the fourth pair, enlarged eight diameters.
" 3 c. Leg of the seventh pair, enlarged eight diameters.
" 4. *Cirolana concharum* Harger ex Stimpson. Last segment of pleon with uropods, enlarged six diameters.
" 4 a. Leg of the first pair, enlarged eight diameters.
" 4 b. Leg of the fourth pair, enlarged eight diameters.
" 4 c. Leg of the seventh pair, enlarged eight diameters.

PLATE III.

Fig. 1. *Aega incisa* Schiödte & Meinert. Dorsal view of specimen from Station 307, enlarged five diameters.
" 2. *Rocinela oculata* Harger. Dorsal view of specimen from Station 305, enlarged six diameters.
" 2 a. Ventral view of same specimen, enlarged six diameters.
" 3. *Rocinela Americana* Schiödte & Meinert. Dorsal view of female, enlarged three diameters.
" 3 a. Ventral view of same specimen, enlarged three diameters.
" 4. *Rocinela Americana* Schiödte & Meinert. Head and first thoracic segment of male, enlarged three diameters.
" 5. *Syscenus infelix* Harger. Dorsal view of male, enlarged one and one half diameters.
" 5 a. Lateral view of same specimen, enlarged one and one half diameters.

PLATE IV.

Fig. 1. *Rocinela oculata* Harger. Leg of the first pair from specimen figured on Plate III., enlarged fifteen diameters.
" 2. *Rocinela Americana* Schiödte & Meinert. Leg of the first pair, enlarged ten diameters.
" 2 a. Leg of the sixth pair of the same, enlarged six diameters.
" 3. *Syscenus infelix* Harger. Inferior view of the head, right antenna removed to show the antennula, enlarged eight diameters.
" 3 a. Left mandible of same, enlarged twenty diameters.
" 3 b. Maxilla of the first or inner pair, enlarged twenty diameters.
" 3 b'. Tip of same, enlarged about seventy-five diameters.
" 3 c. Left maxilliped of same, enlarged twenty diameters.
" 3 d. Leg of the first pair of same, enlarged four diameters.
" 3 e. Leg of the fourth pair of same, enlarged four diameters.
" 3 f. Leg of the sixth pair of same, enlarged four diameters.
" 3 g. Pleopod of the second pair of same, male, enlarged four diameters.
" 3 h. Uropod of same, enlarged four diameters.

PLATE I.

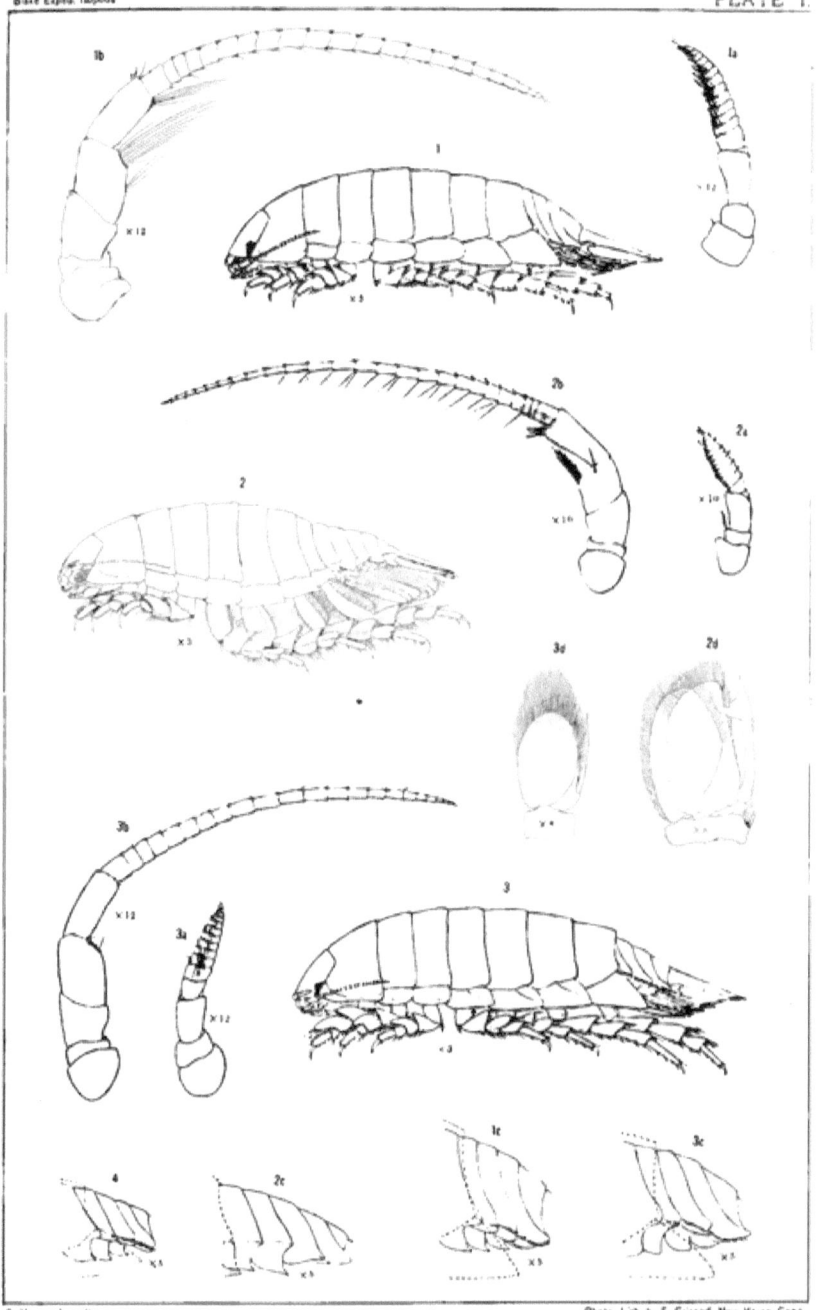

CIROLANA POLITA. C. SPINIPES. C. IMPRESSA. C. CONCHARUM.

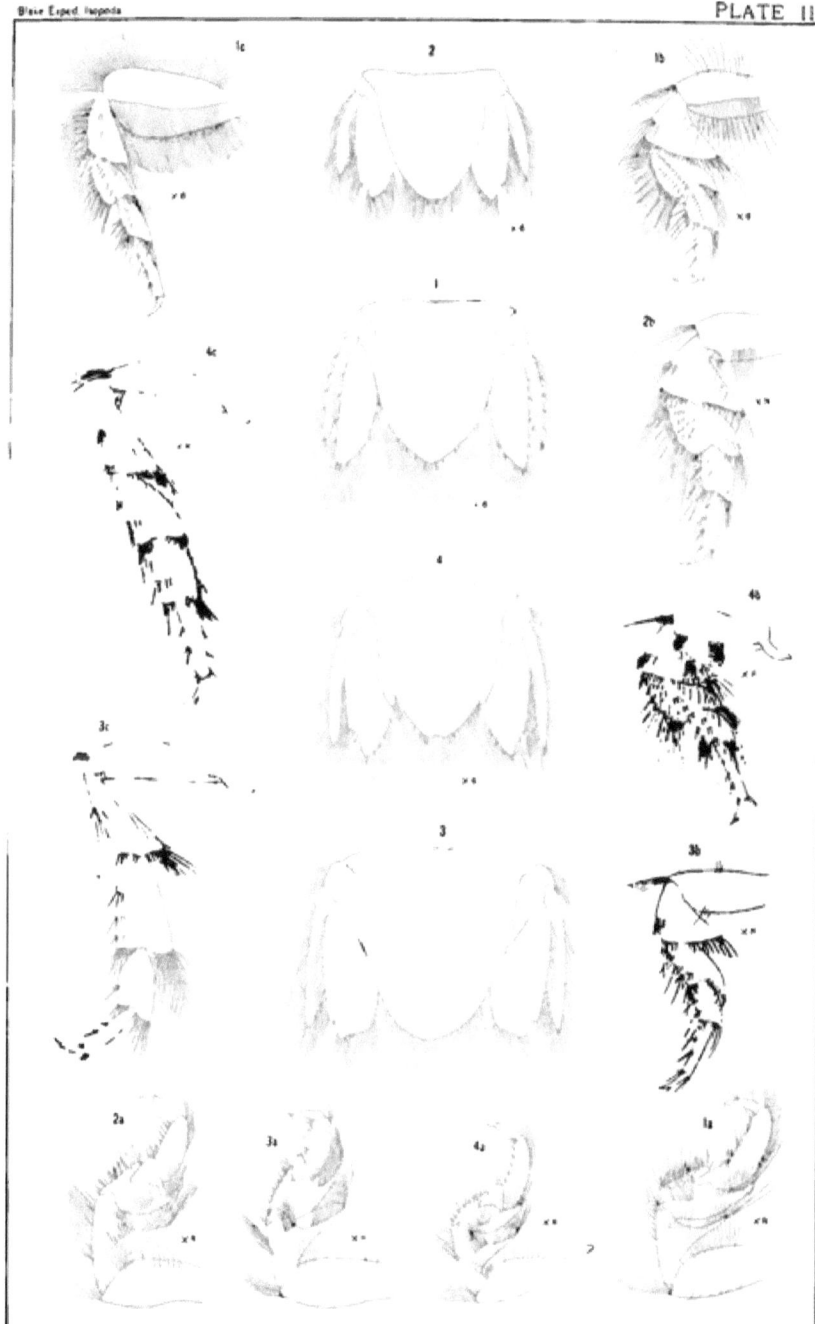

PLATE II.

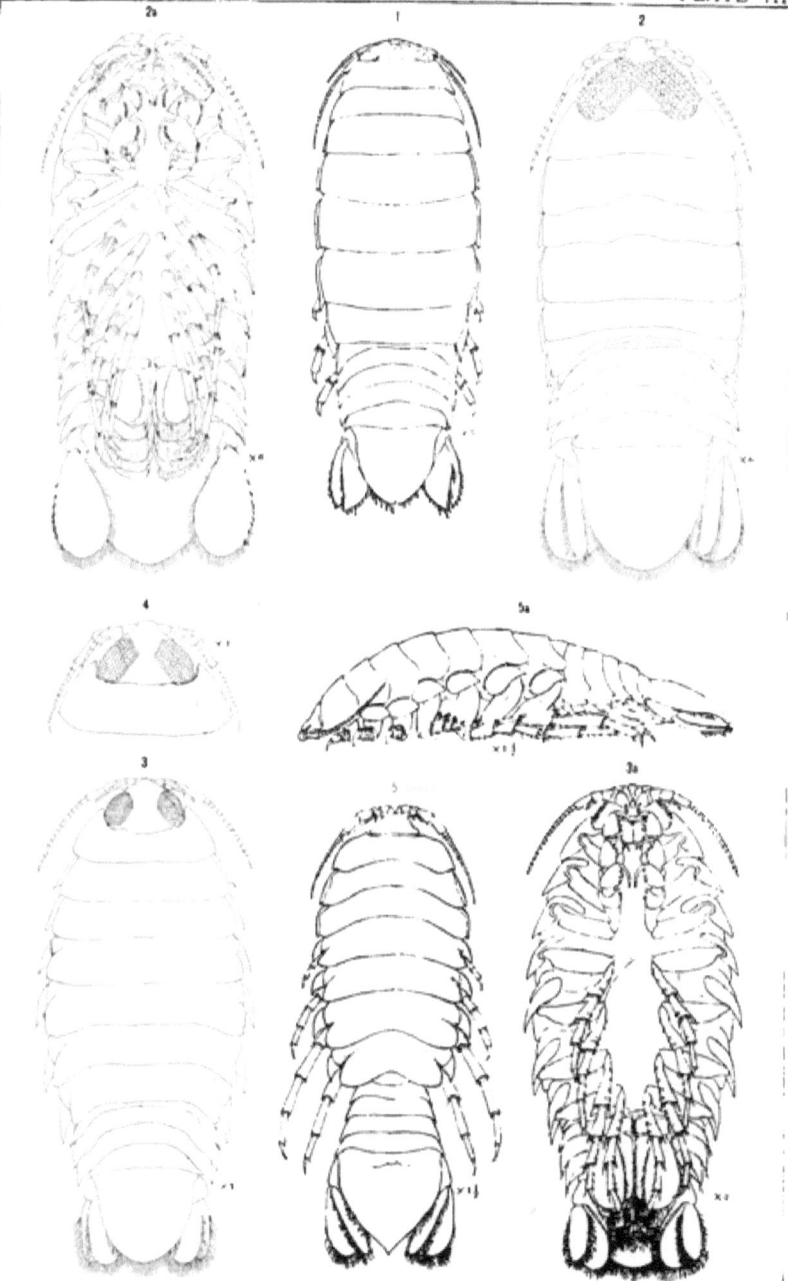

PLATE III.

PLATE IV.

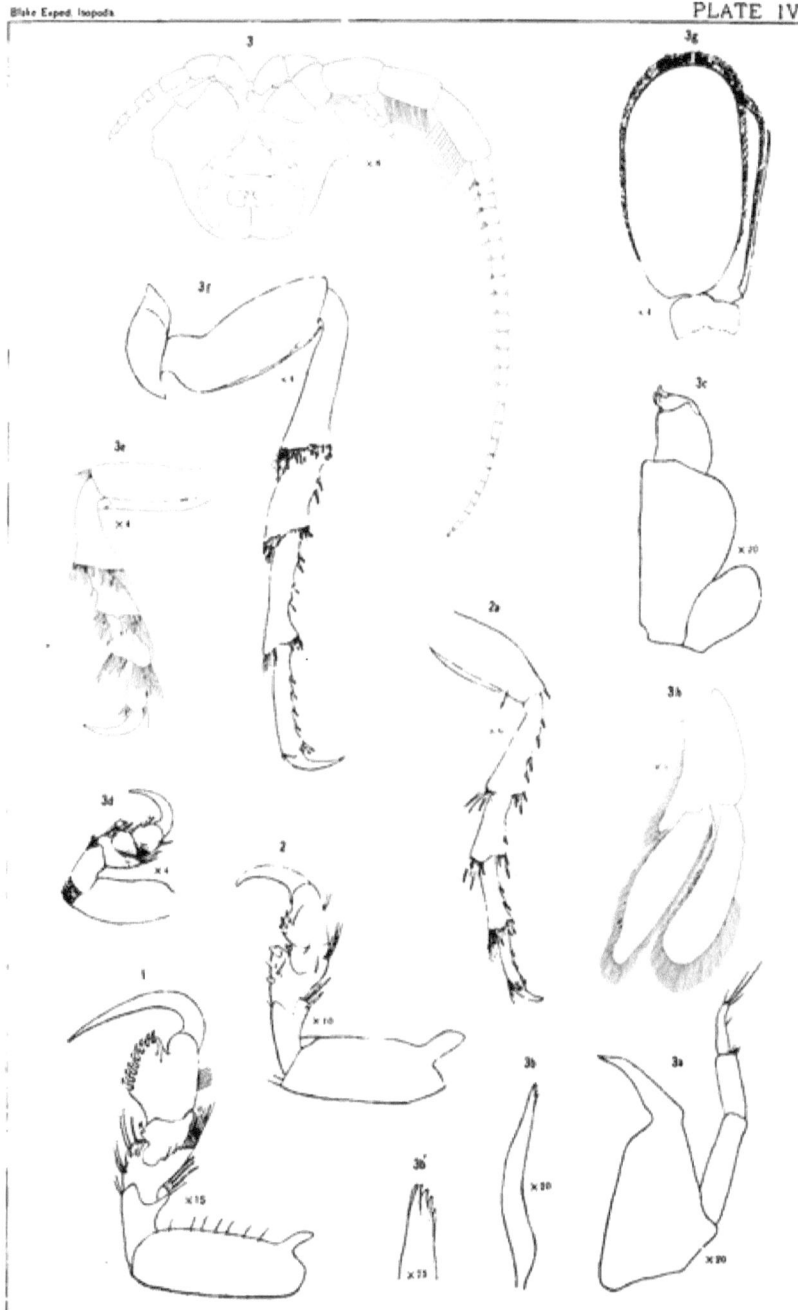

www.ingramcontent.com/pod-product-compliance
Lightning Source LLC
Chambersburg PA
CBHW032136230426
43672CB00011B/2355